The Atlas of African Affairs

The Atlas of African Affairs is divided into five sections dealing with environmental, historical, political and economic issues and with southern Africa. Throughout, the book presents an interdisciplinary, integrated perspective on African affairs.

Most of the chapters deal with continent-wide themes and are illustrated by maps of Africa as a whole drawn to a standardized outline of the same map projection and scale. Other chapters, often by way of example, discuss parts of the continent or individual countries and are illustrated with appropriate maps.

The basic format of integrated text and maps is supplemented by guides to further reading at the end of each section as well as a series of detailed statistical tables at the end of the book.

Ieuan Ll. Griffiths is Reader in Geography at the School of African and Asian Studies at the University of Sussex.

The Atlas of African Affairs

Second edition

Ieuan Ll. Griffiths

London and New York *Witwatersrand University Press*

First published 1984
by Methuen & Co. Ltd
Revised edition 1985
Reprinted by Routledge 1990
Second edition 1994
by Routledge
11 New Fetter Lane, London EC4P 4EE

Simultaneously published in the USA and Canada
by Routledge
29 West 35th Street, New York, NY 10001

Published in the Republic of South Africa
by Witwatersrand University Press
1 Jan Smuts Avenue, Johannesburg 2001, South Africa

© 1984, 1985, 1993 Ieuan Ll. Griffiths

Typeset in Ehrhardt by
Solidus (Bristol) Limited
Printed and bound in Great Britain by
Biddles Ltd, Guildford & King's Lynn

British Library Cataloguing in Publication Data
A catalogue record for this book is available from the British Library

Library of Congress Cataloging in Publication Data

Griffiths, Ieuan Ll., 1934-
    An atlas of African affairs / Ieuan Ll. Griffiths.
    p.   cm.
    First published 1993 in London by Routledge.
    Simultaneously published in the USA and Canada.
    Includes bibliographical references and index.
    Contents: Environmental – Historical – Political – The South.
    ISBN 0-415-05487-7. – ISBN 0-415-05488-5
    1. Africa–Maps.   2. Africa–History–Maps.   3. Africa–Politics
and government,–1960-   4. Africa–Economic conditions–1960-
5. Africa–Ecology.   I. Title.
G2445.G7   1994 ‹G&M›
912.6-dc20   93-13965
                                                                 CIP
                                                                 MAP

ISBN 0-415-05487-7   ISBN 0-415-05488-5 (pbk)

South African ISBN: 1-86814-238-8 (pbk)

Contents

Notes on maps

The maps of continental Africa are drawn on Lambert's Azimuthal Equal-Area Projection based on the Equator and the 20° East Meridian which are represented as straight lines.

Maps showing 'European penetration to 1880' and 'The scramble for Africa' take as a basic source maps in J.S. Keltie (1893) *The partition of Africa*, London, Stanford.

Mineral and mining symbols, wherever shown, are according to the key on the 'Minerals and mining' map (p. 145).

Please note that during the final printing stage of this book, Eritrea achieved its independence.

Maps drawn by Susan Rowland.

Preface

This book is very much the product of teaching and researching for many years in the stimulating inter-disciplinary environment of the School of African and Asian Studies at the University of Sussex. I am grateful to colleagues at Sussex, in Geography and other disciplines, with whom I have taught, written, and discussed over the years. The university library and the library of the Institute of Development Studies at Sussex have been wonderful providers of material and I am indebted to staff in both institutions for their support. My experience of Africa now extends over thirty years and almost as many countries where many have unstintingly helped in many ways.

My particular thanks go to Sue Rowland who has redrawn all the old maps as well as drawing all the new ones. Since the first edition Sue has successfully converted to computer cartography and has achieved the 'impossible' of improving on her previous efforts which won generous acclaim from reviewers.

To the many reviewers of the first edition from around the world my thanks for their encouragement and constructive criticism. The anonymous international referees of the new edition provided many useful suggestions which are gratefully acknowledged. Tristan Palmer and his team at Routledge have given much appreciated support with just about the right amount of stick and carrot!

Final responsibility for errors of fact or judgement is mine but it is good to have had generous assistance along the way from so many.

Ieuan Ll. Griffiths
University of Sussex
1993

Abbreviations

AEF	*Afrique Equatoriale Française*
ANC	African National Congress
AOF	*Afrique Occidentale Françasie*
AWB	*Afrikaner Weerstandsbeweging*
BADEA	Arab Bank for Economic Development in Africa
CAR	Central African Republic
CEAO	*Communaute Economique de l'Afrique de l'Ouest*
CITES	Convention on International Trade in Endangered Species
CKD	completely knocked down
CODESA	Conference for a Democratic South Africa
CSO	Central Selling Organization (Diamonds)
dwt	dead weight tons
EAC	East African Community
ECA	Economic Commission for Africa
ECOWAS	Economic Community of West African States
ELF	Eritrean Liberation Front
EPLF	Eritrean People's Liberation Front
FAO	Food and Agriculture Organization
FNLA	*Frente Nacional de Libertacao de Angola*
FRELIMO	*Frente de Libertacao de Mocambique*
GDP	Gross Domestic Product
GNP	Gross National Product
HDI	Human Development Index
IBRD	International Bank for Reconstruction and Development (World Bank)
ICJ	International Court of Justice (at The Hague)
IMF	International Monetary Fund
ITCZ	inter-tropical convergence zone
LDCs	Less Developed Countries
LNG	liquefied natural gas
MNR	Mozambique National Resistance
MPLA	*Movimento Popular de Libertacao de Angola*
OAU	Organization of African Unity
ODA	Official Development Assistance
OPEC	Organization of Petroleum Exporting Countries
POLISARIO	*Frente Popular para la Liberacion de Saguia el Hamra y Rio de Oro*
PTA	Preferential Trade Area

SACU	Southern African Customs Union
SADCC	Southern Africa Development Co-ordination Conference
SADR	Saharan Arab Democratic Republic
SWAPO	South West Africa People's Organization
TAZARA	Tanzania–Zambia Railway
UAR	United Arab Republic
UDEAC	*L'Union Douaniere et Economique de l'Afrique Centrale*
UDI	unilateral declaration of independence
UN	United Nations
UNCTAD	United Nations Conference on Trade and Development
UNDP	United Nations Development Programme
UNITA	*Uniao Nacional para a Independencia Total de Angola*
UPC	Uganda People's Congress
WHO	World Health Organization
ZANLA	Zimbabwe African National Liberation Army

1 Introduction

Africa is a vast and fascinating continent, home to over 500 million people, living in more than fifty separate states. Africa faces more problems of a life and death nature and has probably undergone more economic and political change in this century than any other continent. Those two propositions are closely linked but the relationship between dramatic changes affecting the way people live on the one hand and life-threatening human problems on the other are complex and indirect rather than simple and direct. It helps to explore the complexities of that relationship within a geographical or spatial framework, whether at the continental scale or the scale of the individual country. In an introductory manner that is what this book seeks to do, using the format of each chapter of text illustrated by one or more specially drawn maps to give spatial focus to the topic under discussion.

Most of the chapters deal with continent-wide themes and are illustrated by maps of Africa as a whole drawn to a standardized outline of the same map projection and scale. Other chapters, often by way of example, taken as their subject parts of the continent or individual countries and are illustrated by appropriate maps. The basic format of integrated text and maps is supplemented by lists of further reading at the end of each section and a series of detailed statistical tables at the end of the book. Occasionally individual chapters themselves incorporate tables of data where this is deemed to be appropriate.

The book is broken down into five basic sections, Environmental, Historical, Political, Economic and the South, titles which are largely self-explanatory. The first deals with the basic physical and human environment of Africa, dealing not only with Africa's geographical position in the world, its physique, climate, soils and vegetation but also with the direct impact of those elements on human occupance, for example, in the form of drought and disease. Human aspects of the basic environment of Africa include population, ethnic divisions described through language groups, literacy and human well-being as measured by the Human Development Index.

The Historical section discusses Africa as the cradle of humankind through to the point in the latter part of this century when a majority of its people broke away from the colonialism imposed by Europe. A firm grasp of the historical background is essential for an understanding of the contemporary problems which confront modern Africa. In this context again, a fuller and better understanding is gained if that background is explored with the use of detailed maps.

The Political section begins with identifying the states of modern Africa, their shapes and sizes, their location in respect to access to the sea, their boundaries and their international relationships within the continent. At independence Africa was divided into over fifty sovereign states, a recipe for neo-colonialism, political impotence and economic dependence. Attempts to bring states together in formal political unions are surveyed and illustrated by the case of Libya which has unsuccessfully attempted unions with three of its neighbours in the past twenty years. The political instability of Africa is typified by the high number of *coups d'état* experienced and the incidence of military governments. Uganda, Zaire and Nigeria have all suffered from different forms of this type of political instability, have suffered civil war and all still have military leaders. Devastating civil war now characterizes Somalia, all the more tragic as it followed hard on the heels of another expression of political instability, Somali irredentism. The drive to unite people of the same ethno-linguistic group into one political entity is a phenomenon rarely encountered in Africa, but was, and potentially will be again, a root cause of international instability in the Horn of Africa. The Somali situation is exacerbated by the interest that was shown in the Horn of Africa by the superpowers during the cold war. Deemed 'strategic' in global terms, the Horn, and Angola, became killing fields of superpower rivalry played out by proxies. Of the old colonial powers France exerts the most direct influence in Africa but of greater, if more geographically limited, impact is the African imperialism practised by Ethiopia and Morocco towards neighbouring territories.

The Economic section explores the causes of and possible remedies for the economic ills of Africa. Poverty is all too common in Africa and too many people are exposed to its deprivations. African economies still have large rural components and the dualism of modern and traditional persists. Famine dominates news out of Africa and its causes are many. In part it is a legacy of colonialism and a world economic system in which Africa's role is as supplier of raw materials and cash crops. Wars are an immediate cause of poverty and, with natural disasters, make millions of Africans into refugees, unable to grow life-sustaining crops or even to work. Development is often more than offset by population growth and neo-Malthusian theses again flourish to explain the phenomenon. The desperate poverty of Africa, which leaves millions at the brink of starvation, is met in part by international aid of two kinds, emergency supplies and longer term development aid. Many national economies in Africa are deeply in debt to Western banks and governments which, through the International Monetary Fund (IMF) and the World Bank (IBRD), increasingly have imposed stringent conditions on African governments seeking help with their beleaguered economies. The contrasting development paths taken by Ghana and the Ivory Coast illustrate different strategies to beat poverty and dependence. Mineral development is of crucial importance in Africa but it is a path fraught with difficulties as the case of Zambia exemplifies. The modern economic sector, as represented by manufacturing, is the least developed in Africa generally, but the infiltration of Western capital increases. Other Western influence is seen in the development of tourism, which sometimes conflicts with other economic development as, for example, when space for wildlife, a major tourist asset, is squeezed by demands to use that space for rural people to graze their cattle and otherwise earn a living. Geographical differences in the intensity of African development lead to migration of people towards jobs, both within countries, from rural to urban, and across borders, in temporary labour migration. The post-independence growth in urbaniz- ation is a major headache for many African countries. It is exacerbated when much of the urban inflow is directed towards the capital city. As part of the colonial inheritance most African capitals are ports and, being peripheral in location, skew national economies and hamper economic development. Some countries have built new capital cities which form part of a new programme of infrastructural development necessary to underpin a modern economy. Transport and inanimate energy, including the realizing of some of the enormous potential that exists in Africa for hydroelectric power, are key elements in the long-term modernization of African economies. But most important of all is the realization that, in face of the political balkanization of Africa, there is a very real need to create larger economic units in Africa, in the form of groups of countries, in order to begin to enjoy the economies of scale necessary for accelerating economic development.

European colonialism, in its widest sense, is not yet over in Africa. South Africa is still under white minority rule. Attempts by the South African regime in the 1980s to prolong the apartheid state by destabilization of the neigh- bouring countries took a grievous toll on the economies and infrastructures of the front line states. It was not until 1991 that attempts were first made seriously to solve the crisis of apartheid by peaceful means when Mandela was released and the African National Congress (ANC) unbanned. Negotiations within South Africa to end minority rule are slow and difficult, with the spectre of all-out violence increasing with every individual flare-up. The outcome of the talks will have an enormous effect on the future not only of South Africa but also of all the sub-continental region. To resolve the situation peacefully will be a great achievement but even then the future will hold great challenge.

Such is the ground covered by this book. It is both coherent and cohesive, which is not to claim that it is comprehensive in any way. Every reader will think of additional, even alternative, topics, of different emphases and inevitably of different examples. There are omissions and overlaps, so that a certain amount of cross-referencing is necessary. For example, Chapter 9 'Population' and Chapter 46 'Development and population' complement each other and are supplemented by the detailed population statistics at the end of the book.

The work is unashamedly idiosyncratic, born of long personal experience of Africa and African studies. It makes no claim to be 'objective', whatever that may be, though great care has been taken to be as accurate and up to date as possible with facts. Any work of this kind is bound to be subjective, based as it is on individual experiences and a particular background which together affect the selection of topics as well as the treatment of each topic and the opinions expressed.

A Environmental

2 Africa: barrier peninsula

Africa is completely surrounded by water except where it borders on Asia. The Egyptian king Necho was the first to establish this fact. After he desisted from trying to dig the canal that extends from the Nile to the Arabian Gulf, he sent some Phoenicians in ships with orders to sail back into the Mediterranean Sea by passing through the Pillars of Hercules and so return to Egypt. The Phoenicians left Egypt by way of the Red Sea and sailed into the southern ocean. When autumn came, they went ashore, wherever in Africa they were, to sow grain and await the harvest. On reaping the grain they again set sail and thus after two years had passed they rounded the Pillars of Hercules and in the third year reached Egypt. They told a tale that I do not believe, though others may, that in sailing along the African coast they had the sun on their right hand.

(Herodotus *c.* 440 BC)

Herodotus' throw-away line at the end is, of course, the very basis for believing the whole story. The canal was duly completed from Suez to the Nile in 521 BC by Darius the Persian. He marked its course with three rose granite *Stelae* which record: 'this canal was dug as I [Darius] commanded, and ships passed through this canal to Persia as was my will.'

So the ancient world came to grips with the basic geographical problem posed by Africa as a vast barrier peninsula joined to the Eurasian land-mass by the 100 mile (160 km) wide isthmus of Suez. Modern solutions to the problem are basically the same as the ancient. In AD 1498 Vasco da Gama circumnavigated Africa from west to east, and in AD 1869 a new canal, this time direct to the Mediterranean Sea from Suez, was at last completed by Ferdinand de Lesseps.

The African barrier is more formidable than the narrow isthmus of Suez or the 10,000 mile (16,000 km) coastline in themselves suggest. The Sahara desert, some 1500 miles (2400 km) across, completely spans the great northern width of Africa, severely limiting overland and, for centuries, even coastal communication. From the seventh century AD Islam spread across northern Africa bringing extensive cultural change and forming another barrier to wider contacts. The 'barrier' role applies to Africa today. Supertankers carrying oil from the Persian Gulf are too big to go through even the enlarged Suez canal and have to take the Cape sea route. During recent (post-1956) enforced closures of the canal *all* east–west maritime trade has had to make that long haul.

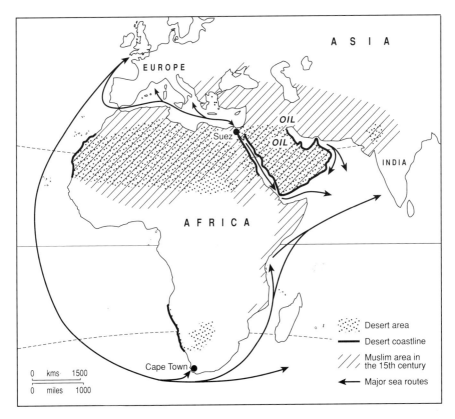

For black Africa the barriers have meant isolation that has never been complete but has been serious enough, for example, to account for the absence of the wheel. This geographical isolation, in part at least, may have been responsible for the development of apartheid in South Africa directly against the contemporary trend in the West towards decolonization and liberalization. But isolation can have its advantages. For example, in forty years of threatened nuclear catastrophe Africa's physical isolation protected it from immediate risk apart from a possible South African capability. In the last decade of the superpower cold war it was suggested by a leading African political scientist that a way to strengthen Africa's political position in the world would be for one or more of the richer black African states to acquire a nuclear capacity. As the possibility of superpower nuclear confrontation fades and South Africa edges towards majority rule, a permanent nuclear-free zone south of the Sahara, including South Africa, is much more desirable and seems readily achievable.

3 The physique of Africa

Africa straddles the equator, extending almost as far south as it does north. This simple geographical fact is of enormous significance as it is the basis for understanding the symmetrical distribution of African climates, vegetation and peoples on either side of the equator. Despite their very different shapes and sizes, the south is largely a mirror image of the north. The Kalahari matches the Sahara, the Karroo, the Maghreb, and the Cape the Mediterranean littoral. At the centre is the equatorial forest of the Zaire (Congo) basin. To the north and north-east Africa is separated from Eurasia by narrow seas but is also joined to it by the isthmus of Suez. The adjacent location of Arabia, and beyond that Persia, means that north-eastern Africa is arid almost to the equator, an important exception to the symmetrical pattern.

Africa comprises a single tectonic plate; though some would differentiate the area east of the Rift Valley. Almost the entire continent is a geologically stable land-mass of pre-Cambrian basement rocks partly overlain by later sedimentary cover. In the extreme south-west corner the Cape Fold Mountains are of Hercynian age, in the extreme north-west the Atlas Fold Mountains are of the Alpine orogeny. Elsewhere the stability is broken only by the Rift Valley system.

In contrast to Europe the continent of Africa has a remarkably smooth outline. Its coastline is short in relation to its area and there are few major inlets or peninsulas. On a smaller scale there is a marked absence of natural harbours. The continental shelf of Africa, again in contrast to Europe, is almost uniformly narrow. The major exception is the Agulhas Bank off the southernmost tip of Africa which extends some 200 miles (320 km) off shore. The absence of a wide continental shelf limits fishing opportunities and reduces the chances of finding oil fields. Africa has relatively few offshore islands and most are small and of volcanic origin. The major exception is Madagascar, which is the world's fourth largest subcontinental island.

The ocean currents off the African coast are influenced by the continental straddling of the equator. On the east coast the westward flowing North-Equatorial Current of the Indian Ocean divides to flow northwards as the Monsoon Drift and southwards as the Mozambique Current. The Monsoon Drift flows northwards in the northern summer but is reversed in the northern winter, historically a major factor in trading links between east Africa and Arabia, the Gulf and India. The Mozambique Current sweeps down the coast of south-eastern Africa as a swift warm current. On the west coast the currents flow towards the equator as the Canaries Current and the Guinea Current

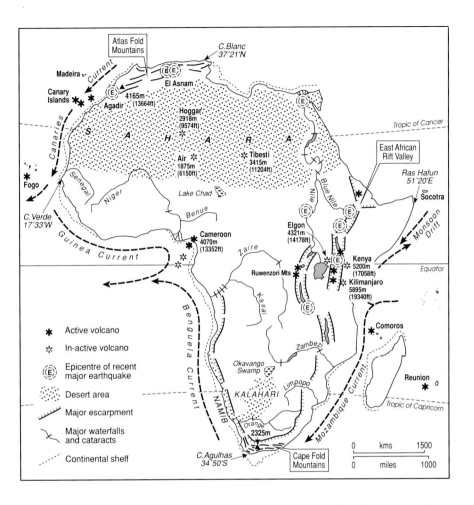

from the north and the Benguela Current from the south. They are cold currents which flow for hundreds of miles along hot desert coasts well known for good fishing and for hazardous fogs caused by this juxtaposition.

Africa is a continent of wide horizons on broad, flat plateau surfaces. Plains cover thousands of square miles, stretching away, seemingly endless, in a largely featureless landscape. The plateau consists of a number of vast, shallow basins separated often by barely discernible watersheds. Occasionally there are mountainous tracts of considerable height, as in the Tibesti, Aïr and Hoggar mountains of the Sahara. In southern Africa, from the mouth of the Zaire (Congo) in the west to Malawi in the east, the Great Escarpment, which is outward-facing and parallel with the coast, in places rises to over 10,000 feet

(3000 m) and is a formidable obstacle to transport development from the sea.

The Cape Fold Mountains are wrapped, in a series of parallel ranges, around the south-west corner of the continent. They are aligned north–south along the west coast for about 150 miles (240 km) then swing through 90 degrees to run west–east along the south coast for over 600 miles (900 km). The ranges are steep and high, reaching 7632 feet (2325 m), with literally dozens of peaks of over 5000 feet (1500 m). The Cape Mountains make for spectacular scenery especially near the coast, as in the Cape Peninsula, but also in the narrow gorges (*poorts*) cut through the ranges. Being near to the coast and roughly parallel with it throughout their length, the Cape Fold Mountains have been a most effective barrier to penetration of the continental interior both in the past by humans and up to the present by rain-bearing winds.

The Atlas Mountains occupy the north-western corner of Africa and are an extension of the Alpine system of Europe. Consisting of a number of parallel ranges with intermontane plateaux and valleys they extend in a belt up to 200 miles (320 km) wide from Tunisia to southern Morocco, a distance of 1400 miles (2250 km). Seldom above 5000 feet (1500 km) in Tunisia, there are two main ranges in Algeria divided by a plateau. In Morocco there are four main ranges, the Rif, Middle, High and Anti Atlas and it is in the High Atlas that the greatest elevation of 13,664 (4165 m) is attained, far higher than the older Cape Mountains. The Atlas Mountains are parallel with the northern coastline of the Maghreb and so cut off the north-western Sahara from maritime influence.

The ancient surfaces of Africa are disrupted on the eastern side of the continent by the great Rift Valley system which extends roughly north–south from the Red Sea to the Zambesi (and northwards along the Red Sea to the Dead Sea and the Jordan valley). From the Red Sea the Rift Valley cuts through the Ethiopian Highlands; in east Africa there are two Rift Valleys, eastern and western, which unite in the southern or Malawi section. The rifts are prominent features in the landscape, great trenches 20–50 miles (30–80 km) wide, with inward-facing walls, sometimes very steep, sometimes stepped, themselves up to 3000 feet (1000 m) high, an unmistakable and unforgettable sight.

The east African rifts have closely associated volcanic and seismic activity and many of the active volcanoes of Africa are to be found here, some towering on the valley sides, some actually on the rift floor. Near the eastern edge of the eastern rift are the two highest mountains in Africa, Kilimanjaro (19,340 ft, 5895 m) and Kenya (17,058 ft, 5200 m), both inactive volcanoes. The Rift Valley areas experience frequent earth tremors and occasional large earthquakes including the largest in Africa this century, near the southern end of Lake

Tanganyika in the western rift. Other earthquake-prone areas in Africa include the Atlas Mountains where recently there has been considerable loss of life at Agadir and twice at El Asnam. Africa's largest active volcano is Mount Cameroon (13,352 ft, 4070 m) which is part of a chain of volcanoes which stretches far out to sea. It is recorded that when the Portuguese first sailed into these waters in the fifteenth century and saw on either side towering volcanoes belching smoke and fire they turned and fled, convinced that they had found the very Gates of Hell. Africa's other active volcanoes are on the offshore islands, notably the Canaries, where, in a less religious age, they are now a major tourist attraction well worth a visit.

Also associated with the Rift Valley system are the great lakes of east Africa. In the Rift Valley floors they are characteristically long, narrow and sometimes of great depth; on the plateau between the rifts lies the relatively broad and shallow Lake Victoria, the third largest lake in the world. In the Lake Victoria basin is the long-sought-after source of the Nile, reputedly the longest river in the world (4150 miles, 6640 km). While almost one-third of Africa is desert, another third is drained by five great rivers, the Nile, Zaire (Congo), Niger, Zambesi and Orange. Because of the plateau basin structure of the Africa interior none of the major rivers is navigable inland from the sea for any great distance, in contrast to the great rivers of the Americas and Europe. The Nile is navigable 960 miles (1500 km) to Aswan, the Zaire 150 miles (240 km) to Matadi, the Niger 120 miles (200 km) to Onitsha all the year round, the Zambesi is open only to shallow-draught boats to Tete 300 miles (500 km) from the mouth, while the Orange is simply not navigable (except in the fertile imagination of Jules Verne). The Nile, Zaire and Niger do have long navigable interior stretches but they are also broken by further falls and cataracts. The Zaire has the most extensive inland waterway system with the greatest stretch from Kinshasa to Kisangani, about 1000 miles (1600 km) in Conrad's 'heart of darkness'. The absence of navigable waterways from the sea kept aliens, such as explorers, traders and colonizers, at bay and it is still an impediment to economic development; however, such disadvantage is balanced by the enormous hydroelectricity potential that is now being realized on all of the major rivers at points where they plunge over the basin rims and escarpments.

The unique characteristics of the physique of Africa have done much to shape the lives of its peoples and to provide it with enormous potential for future development.

4 Sunshine and storm

Rainfall is the climatic factor of greatest significance in Africa. Most of the continent has a small annual range of temperature, and wind is also much less of a feature than in temperate latitudes. Africa extends little beyond 35 degrees of latitude from the equator. This limits the range of African climates and also means that the basic movement of air over most of the continent is towards the equator or, more accurately, towards the inter-tropical convergence zone (ITCZ). The actual position of the ITCZ shifts with the seasonal movement of the sun across the tropics. In July the ITCZ normally lies across north Africa along the southern edge of the Sahara; in January it normally skirts the west African coast, snakes along the northern and eastern margins of the Zaire (Congo) basin and thence over Madagascar. The position of the ITCZ influences the distribution of climatic zones, which assume some symmetry about the equator as Africa extends almost as far south as it does north of the line. That symmetry is distorted by the effect of the adjacent Eurasian land-mass and by highland areas within Africa itself.

The climatic zones of Africa are shown according to Thornwaite's classification. Largely because only the two warmest of the six 'temperature efficiency' classes are present, Africa has only nineteen different climatic zones, most of which are found in more than one part of the continent.

Thornthwaite's classification

1	Precipitation effectiveness	2	Temperature efficiency	3	Seasonal rainfall
A	Wet	A′	Tropical	v	adequate all seasons
B	Humid	B′	Mesothermal	s	deficient in summer
C	Sub-humid	*C′	Microthermal	w	deficient in winter
D	Semi-arid	*D′	Taiga	d	deficient all seasons
E	Arid	*E′	Tundra		
		*F′	Frost		

*Not represented in Africa

The wet tropical climates (AA′v) in Africa are limited to the coastal strips of Sierra Leone/Liberia, the Nigeria/Cameroon border, and eastern Madagascar. These areas experience rain at all seasons with an annual average of up to 200 inches (5000 mm), and mean annual temperatures of

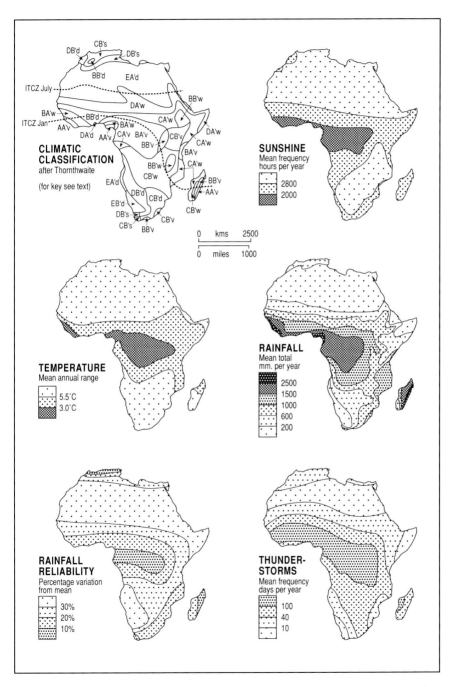

CLIMATIC CLASSIFICATION
after Thornthwaite

(for key see text)

CB's
DB'd
DB's
BB'd
EA'd
ITCZ July
BB'w
DA'w
BA'w
BB'd
CA'w
ITCZ Jan
BA'w
AA'v
DA'd
CA'v
BA'v
CB'v
DA'w
AA'v
BB'v
CA'w
BA'v
EA'd
CA'w
BB'v
CB'w
AA'v
DB'd
CB'd
EB'd
CB'w
DB's
CB's
BB'v
CB'v

0   kms   2500
0   miles   1000

SUNSHINE
Mean frequency
hours per year

2800
2000

TEMPERATURE
Mean annual range

5.5˚C
3.0˚C

RAINFALL
Mean total
mm. per year

2500
1500
1000
600
200

RAINFALL RELIABILITY
Percentage variation
from mean

30%
20%
10%

THUNDER-STORMS
Mean frequency
days per year

100
40
10

about 79°F (26°C) with a very small annual range of temperature of only 4°F (2°C).

Inland, rainfall decreases as a drier winter season (BA'w) develops. The Zaire (Congo) basin is surprisingly dry for an equatorial lowland with a seasonally well-distributed mean annual rainfall of 40–80 inches (1000–2000 mm), mean annual temperatures of 68°–77°F (20°–25°C), low annual temperature range, and high humidity (BA'v). A similar climate is found on the Kenyan and north Tanzanian coasts. The mountains of Rwanda, Burundi and the Ruwenzoris in East Africa, the mountain core of Madagascar and the Tsitsikama mountains on the southern coast of the Cape Province of South Africa all have a similar climate, modified by elevation and latitude (BB'v), with an annual rainfall of up to 100 inches (2500 mm) but lower mean annual temperatures of about 60°F (16°C). The Ethiopian highlands and the highlands of south-western Tanzania and Malawi also have a similar climate but with a dry winter season (BB'w), lower annual rainfall totals and lower mean annual temperatures with frost experienced at higher altitudes.

The sub-humid climates of Africa mainly extend in an arc around the wet and humid equatorial core, but are also found beyond the arid zones at the northern and southern extremities of the continent. The sub-humid, tropical dry winter season belt (CA'w) stretches across Africa from Senegal to Somalia and also appears in northern Mozambique and Madagascar. Annual rainfall is typically 30–40 inches (750–1000 mm), and mean annual temperatures are higher than in the humid zone. Further from the equator the dry winter season becomes longer and more pronounced. In east Africa both elevation and the local influence of Lake Victoria modify the climate, with lower mean annual temperatures and a less marked dry season (CB'v). A similar climate is experienced in Natal. The interior of southern Africa is similar to, but cooler than, the Sudanic belt (CB'w), and over much of southern Africa, even in normal years, rainfall is deficient in all seasons (CB'd). In the western Cape of South Africa and the coastal Maghreb of North Africa a Mediterranean-type climate of warm, dry summers and cool, wet winters (CB's) is experienced. Annual rainfall here is 30 inches (750 mm) or less, about two-thirds of which falls in the four winter months.

The semi-arid climates border the Sahara, hence the same Sahel (border) for the narrow belt of prolonged dry winter season climate (DA'w) which extends right across northern Africa. This belt has an annual rainfall of under 20 inches (500 mm), and high temperatures with a large diurnal range. In southern Africa the semi-desert, which takes in much of the Kalahari, is cooler and is deficient of rain in all seasons (DB'd). Between the deserts and the Mediterranean zones, areas of semi-arid climate are found with summer rainfall deficiency (DB's).

The arid zone of the Sahara desert (EA'd) covers about 30 per cent of the area of the continent. It is dry with extremely low, highly erratic rainfall; extremely high sunshine values of up to 98 per cent; very high temperatures, including the highest ever recorded anywhere in the world at Azizia, Libya (136.4°F, 57.7°C); and a very large diurnal range of temperature. The counterpart in southern Africa, the narrow Namib desert, extends along the coast of Namibia. Its southern extension is rather cooler (EB'd) because of latitude and coastal fogs caused by the cold Benguela Current.

The distribution of African climatic types shows the essential onion-like pattern of climatic zones: from the cloudy, warm, wet, monotonously uniform, thunderstormed core through, layer upon layer, to the sunny, hot, dry desert skin.

Africa really is the continent of sunshine and storm. In the eastern Sahara a wide area experiences in excess of 4000 hours of sunshine a year (over 91 per cent of possible sunshine); in Kampala there are, on average, 242 days a year with thunderstorms. Who can forget stepping on to the airport tarmac at Entebbe to feel the intense heat of the equatorial sun and then, in the Kampala afternoon, the lashing rain, flashing lightning and frightening thunder of the suddenly unleashed tropical storm, followed by the cool, fragrant evening on Makerere Hill, with the night sky illuminated by sheet-lightning? Or, when sailing in air-conditioned comfort up the Red Sea, the literally breath-taking blast of oven-like heat as one goes on deck? Or driving through the pounding hailstones of Kericho, anxious for the haven of the Tea Hotel? Or the Berg wind in February searing down on Durban at 104°F (40°C)? Or waiting for a few hot, humid hours in the *old* customs shed at Lagos airport? Or the old caretaker of the rest-house in up-country Gambia complaining of the pre-Harmattan 'cold' on a bright, sunny and warm November morning? Such are the lasting impressions of African climate.

Climatic conditions in many parts of Africa are trying for humans. The monotony of heat, humidity and daily 'climatic' regime in humid, tropical areas is profoundly enervating. Great stress is imposed by such oppressive heat as is found along the southern Red Sea coast and the Somali interior. Wide areas suffer from low and erratic rainfall, often triggering crippling drought and the human disaster of famine. In contrast many parts of Africa are periodically subjected to torrential rain, flood and devastation. Then there are the indirect effects of climate, namely disease and pestilence. On the bright side, however, the sunshine, warmth and refreshing rains make so much of Africa a delightful physical environment.

5 Soils

The soils of Africa most favourable to human occupance are found in the major river valleys, such as the Nile and the Niger, whose rich alluvial deposits have been worked for millennia and have long cradled flourishing civilizations based on cultivation. With few exceptions, elsewhere African soils are difficult. Natural fertility has to be carefully husbanded as the basic raw materials and the harsh climatic environment conspire against easy human progress.

African soils are highly varied and on a continental scale even the most descriptive maps have vast areas marked 'undifferentiated' or are a mosaic of difficult-to-generalize detail. The limitations of a small-scale map of African soils are almost total. The most one can hope to do is to draw attention to the close relationship between climate and soils by trying to show that the broad pattern of soil distribution bears some resemblance to generalized climatic zones.

The humid tropical areas have deep soils built up by the intense activity of biological and chemical processes which are stimulated by heat and moisture. The soils are protected against erosion by the thick forest cover itself, the rapid decomposition (and renewal) of which adds a steady supply of organic material to the upper soil layers. But this process is negated by the high rainfall which leaches out most of the plant nutrients leaving low-fertility, difficult soils, often with a hard pan formed of iron or aluminium oxides. These difficult soils are known as *latosols* or *ferrosols*, the best known of which is laterite. There are some limited areas of better soils in the wet tropics, as in southern Nigeria. Much needs to be learned about tropical soils in order to use them to best advantage. Massive problems of soil erosion can occur where the forest cover has been cleared on a large scale.

Between the wet tropics and the deserts, transitional soils match the transitional climates and vegetations. Nearest to *latosols* are *luvisols*, which are well-developed soils, locally rich in plant nutrients but often with iron-oxide hard pan making agriculture difficult. Towards the deserts, *arensols* predominate; these are sandy soils, often deep but low in humus content and not very fertile. They give way in turn to *xerosols* which have a very low humus content over unproductive sands and gravels.

In south-western Africa the Kalahari sands extend in a great swathe through Botswana, Namibia, western Zambia, Angola and into Zaire. Soils in this area are arensols, deep but low in humus content and easily broken down into a not very productive sandy tilth. Riding very wide-tyred bicycles in

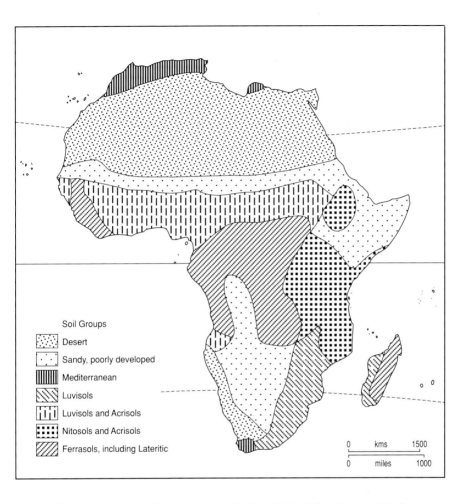

Soil Groups

::::::	Desert
· · ·	Sandy, poorly developed
‖‖‖‖	Mediterranean
⧄	Luvisols
⫼	Luvisols and Acrisols
▪▪▪▪	Nitosols and Acrisols
⧄	Ferrasols, including Lateritic

0	kms	1500
0	miles	1000

Ovamboland or driving low-slung cars in Zambia's Kafue National Park are hazardous occupations in this sand and, once experienced, are ever to be avoided. To eke a living out of such soils is extremely difficult. Even more difficult are the true desert soils, xerosols, *yermosols* and simply blown sand and bare rock.

The soils of Africa are a poor resource which is likely to deteriorate dramatically under pressure from increasing populations. Soil erosion is already a widespread problem and, given that the vast majority of Africans live directly off the land, the consequences of further deterioration could be serious to the point of being catastrophic.

6 Vegetation

The familiar distributional pattern of concentric arcs around the Zaire (Congo) basin derives mainly from the close relationship between natural vegetation and climate, although there are some important local variations owing to soils, drainage and elevation. There are few areas of Africa where 'natural' vegetation has not been modified by human activity: cultivation, herding and hunting. Rapid population growth is dramatically increasing human impact on the vegetational environment.

The tropical rain forest is developed on lowlands with year-round precipitation. It covers much of the Zaire (Congo) basin, extending in long gallery fingers far up tributary valleys and intermittently along the west African coast where the forest belt is widest in Sierra Leone. Forest also extends in a very narrow belt along the east coast south from the equator. In Madagascar there is a rather different form of rain forest with several species not known on the African mainland. The forest is florally rich with numerous species of trees, shrubs, ferns and mosses. The forest floor is dark under a double or triple canopy of trees. Stands of individual species are rare, so commercial exploitation before the advent of the chain-saw was difficult.

Human activities are destroying vast areas of tropical rain forest in Africa. Clearing for agriculture, fuel wood, charcoal and, above all, commercially valuable hard-wood timber has become all too easy. The effects are harmful especially where no replanting takes place. Soil erosion, the destruction of entire plant species and the effects on local and world climates are matters of serious scientific concern.

The margin of the rain forest coincides with the development of a dry winter season. Such areas are a mosaic of forest and savanna. Further away from the forest, where the dry season is more developed, the woodland thins out and more drought-resistant tree species are found. The savanna, the land of big game, is disease-ridden and in general does not support very high densities of human population.

Further still from the equator, the savanna degenerates towards desert. This Sahelian environment ranges from thorn-wooded grassland to tussocky grasses with large patches of bare earth between. The semi-arid areas are increasingly over-populated by humans and animals who both take their toll on the environment. As pasture is destroyed by over-grazing so the desert advances, further restricting populations and increasing densities in a vicious circle of desertification.

Beyond the deserts is the Mediterranean-type vegetation of north and

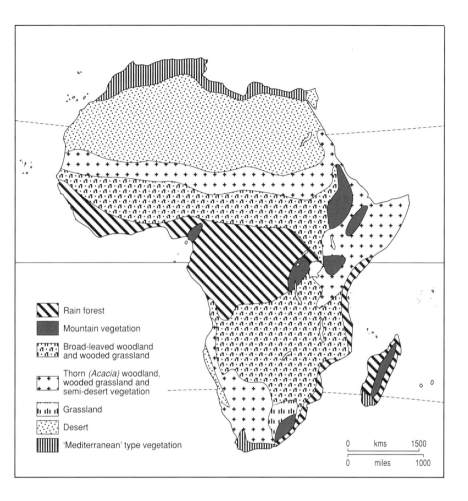

Rain forest

Mountain vegetation

Broad-leaved woodland and wooded grassland

Thorn (Acacia) woodland, wooded grassland and semi-desert vegetation

Grassland

Desert

'Mediterranean' type vegetation

| 0 | kms | 1500 |
| 0 | miles | 1000 |

south, *maquis*-like vegetation capable of withstanding the warm dry summers. In the Cape the flora is unique, but distinctive and indigenous plants such as the protea have to fight for survival against such exotics as the wattle and the blue gum which were introduced during colonial times in what were construed as the interests of human progress.

Conservation in Africa needs to ensure that the often appalling immediate pressures do not result in short-term solutions which are destructive in the longer term. There are too many new Sahelian water boreholes which solve problems of immediate water supply only to lead quickly to the destruction of all vegetation within a wide radius.

7 Drought

Drought is endemic in the semi-arid desert fringes of Africa. The Saharan margins or the Sahel, the lands surrounding the Ethiopian highlands, and large parts of southern Africa have all suffered severe cyclical drought in recent years. These are areas where, at best, humans make a precarious living. In the early 1970s there was severe drought throughout the Sahel and Ethiopia. That crisis continued until after the return of normal rains in late 1974. Since then, rainfall has been below average levels though not at crisis point. In 1980 drought swept through lands marginal to the Ethiopian highlands. The Karamoja district of north-eastern Uganda and the Ogaden in Ethiopia and Somalia were among the worst hit areas. In 1983–5 northern Ethiopia was devastated by severe drought, and in 1991–2 much of Somalia. In southern Africa, too, there have been severe cyclical droughts; in 1983, and again in 1992, large parts of South Africa, Botswana, Zimbabwe and Zambia were affected.

Rainfall in the marginal lands is not only low but also unreliable. Average rainfall is less than 24 inches (600 mm) per annum, falling in a single, short, rainy season which, if deficient, has devastating effects. Rainfall is unpredictable in amount and timing; the smaller the average rainfall the less the reliability. In the early 1970s, for example, the drier lands in the Sahel were worse affected than the wetter; the nearer the Sahara the greater the deficiency against the local rainfall norm. High temperatures and sunshine rates severely reduce rainfall effectiveness, so that an annual precipitation rate which is the equivalent of that of London is, in the Sahel, classed as semi-arid.

All of the drought-stricken areas are apparently subject to long-term cyclic variations in rainfall. In this century the Sahel has experienced severe droughts at approximately thirty-year intervals. The median years of the dry periods were 1913, 1942 and 1971, when respectively 59 per cent, 79 per cent and 70 per cent of 'normal' rainfall was experienced. In each of these dry periods drought conditions prevailed for a number of successive years. Over the five years 1968–72, rainfall averaged only 81 per cent of the norm, creating a cumulative effect of rainfall deficiency more serious than in the two other dry periods of this century. This drought was preceded by above average rainfall throughout the 1950s and early 1960s. Available climatic statistics are generally inadequate for much of Africa because weather stations are too few in number and the period over which most statistics have been collected is too short.

Because of inadequate data it is difficult to decide whether the observed cycles of drought are 'normal' or whether the dry climates of Africa are

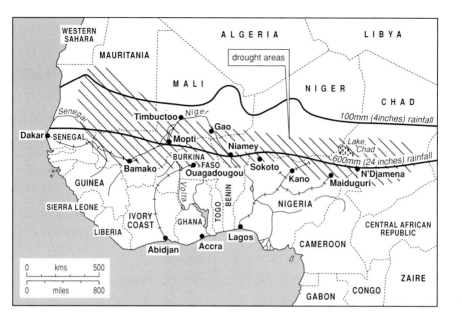

undergoing a secular change. Are these areas actually becoming drier in the long term? If so, why, and what can be done about it? Or is it a matter of scientists and others simply being more aware, for the first time, of what has always prevailed in these arid and semi-arid areas? There is little hard scientific evidence as yet that, long term, these areas are becoming drier. African history is littered with references to drought, even if these are not scientifically recorded. The often devastating human effects which are now witnessed by a world television audience have causes wider than drought itself and forty years ago would have passed largely unnoticed (refer to Chapter 39, Famine).

8 Disease and pestilence

Much of Africa is not a healthy human environment: its climates encourage harmful insects, its water-supply problems make for poor sanitation and hygiene, its poverty causes malnutrition. Many in Africa die from diseases endemic in the continent, but many more are chronically sick from debilitating diseases which seriously impair efficiency and significantly lower the quality of life.

In the 1980s all of tropical Africa plus the densely populated lower Nile valley and delta were malarial. The World Health Organization (WHO) estimates that 250 million people in Africa are exposed to the disease and that about 'one million infants and children die each year'. Malaria is transmitted by some mosquitoes, such as *Anopheles gambiae*, and is combated by draining swamps and pools, the breeding grounds of the mosquito, and by spraying with DDT, which itself is dangerous, destructive, costly and sometimes ineffective.

Yellow fever, a deadly disease which is also transmitted by mosquitoes, such as *Aedes aegypti* and *A. africanus*, has been effectively controlled by vaccination and insecticides. However, in tropical Africa epidemics still occur among populations not recently exposed to the virus. In 1982 eight African states reported infected areas. In the same way, cholera and plague are still endemic to Africa. In 1982 fourteen states reported cholera-infected areas and three, Madagascar, Tanzania and Zimbabwe, plague. Numbers now affected by these particular diseases are small but they do emphasize the difficulty of complete eradication even where immunization is available.

River blindness (*Onchoceriasis*), is transmitted by the small fly *Simulium damnosum*, is widespread and has endemic focuses in most of tropical Africa. In the savanna areas of West Africa alone, WHO estimates that more than one million people are affected. Bilharzia (*Schistosomiasis*) is caused by a blood fluke, hosted by a fresh-water snail, which is picked up by drinking, bathing or washing in infected water. New irrigation schemes have resulted in new focuses of infestation simply because there are more open-water breeding sites available. It is the opinion of WHO that 'human and animal *Trypanosomiasis* (sleeping sickness) is one of the serious obstacles to socio-economic development.' About 35 million Africans are at risk from the disease, which is transmitted by the tsetse fly. Its effect on cattle severely limits the spread of agricultural progress.

Tuberculosis and poliomyelitis, largely eliminated in the developed world, are widely prevalant in Africa. Not least they are diseases of the urban slums,

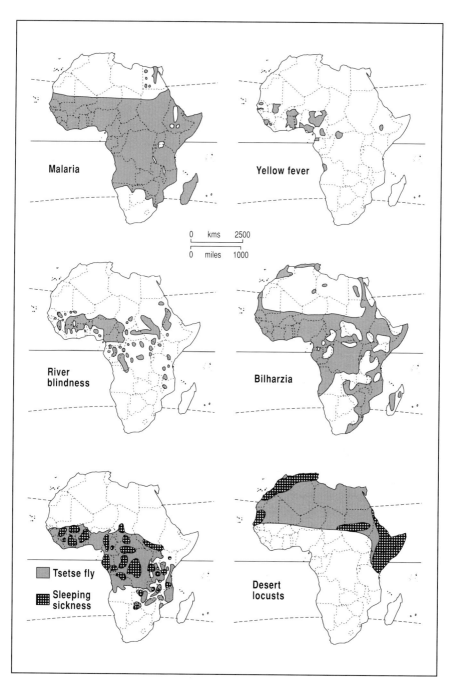

Malaria

Yellow fever

0   kms   2500
0   miles   1000

River
blindness

Bilharzia

Tsetse fly

Sleeping
sickness

Desert
locusts

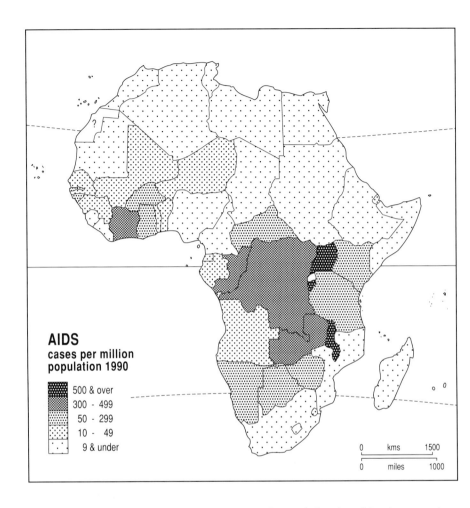

AIDS
cases per million
population 1990

500 & over
300 - 499
50 - 299
10 - 49
9 & under

0   kms   1500

0   miles   1000

where malnutrition, inadequate clothing, inferior shelter (resulting in wet and even cold) and unhygienic conditions are major contributory factors.

In addition to diseases, pests such as the desert locust have made life in rural Africa more precarious since Biblical times. The locust is a major pest in north Africa and the Horn and its depredations often mean the difference between life and death in areas already stricken by drought, as was experienced in Ethiopia in the late 1980s.

From the late 1970s Africa has been plagued by AIDS (Acquired Immune Deficiency Syndrome), a disease which is mainly transmitted sexually via body fluids containing the Human Immunodeficiency Virus (HIV). The virus does not itself kill but destroys the natural defence mechanisms of the human body, thus making lethal what are normally mild and easily picked up infections. More than any other continent, Africa is stricken by the AIDS pandemic and some sources estimate that almost two-thirds of 'the global total of all AIDS cases' are in Africa. In addition, African populations are widely regarded as the most vulnerable to future infection. As with the diseases discussed above, AIDS is likely to thrive in poor societies where medical resources are limited and where social, economic and political instability is likely to undermine efforts to overcome the scourge.

Controversially, many have seen Africa as the place of origin of AIDS but there is no clear evidence that this is the case. The accusation, not surprisingly, has bruised sensitivities and brings with it reinforcement of pre-existing racial prejudices and discrimination. The issue muddies waters and could seriously hamper the search for a means of combating the disease, which would be made easier if the full history of its origin and development were known definitively.

AIDS in the West initially predominated in the homosexual population, but AIDS in Africa spread rapidly and predominantly among the heterosexual population, and is found almost equally in male and female. As a result, more than in any other continent, AIDS in Africa is frequently passed from mother to unborn child. This difference, of course, does not point to one or other place as the original source of the disease. The disease is also spread by contaminated blood and hypodermic needles and in many areas of Africa poor health care facilities worsen the risk of such contamination.

The most affected countries of Africa are all in the tropics with the highest incidence recorded in Uganda, where estimates place the number of HIV positive in the adult population as close to one million. Incidence appears to be highest in south-western Uganda, an area through which a number of armies have rampaged in recent years, though the connection is not fully understood. Other states badly hit by the disease include Zaire, Rwanda, Kenya, Tanzania, Congo, Zambia and Malawi. In West Africa the Ivory Coast and Guinea-Bissau are also badly affected. AIDS is not confined to these countries, however. It is becoming a major problem throughout much of sub-Saharan Africa, in Zimbabwe and in the South African townships.

AIDS apart, progress has been made in disease control in recent years, but much remains to be done before Africa is freed from the debilitating effects of its many endemic diseases. Success in fighting disease, including AIDS, is largely a function of the resources committed and as such is a vital part of the fight against poverty and deprivation. The importance of health education, which requires a basic level general education to be effective, cannot be over-emphasized.

9 Population

The total population of Africa, by United Nations (UN) estimates, probably exceeds 500 million. The continent has an average population density of about 45 persons per square mile (17 per sq. km) compared with that of India of about 550 per square mile (200 per sq. km). In overall terms there appear to be no grounds for concern about African population levels, but population growth rates in several African countries are now among the highest in the world and are undermining economic achievement by exceeding rates of economic growth.

Average figures convey little because the distribution of population is far from uniform. Three vast areas, the Sahara, the Kalahari and Namib deserts, and the tropical rain forest of the Zaire (Congo) basin, support very small populations for obvious reasons. On the other hand, such areas as the lower Nile valley and delta of Egypt and the Mediterranean coastal belt of the Maghreb each have populations of about 40 million at high density. South of the Sahara the greatest concentration of population, over 50 million, is inland from the Gulf of Guinea from Ivory Coast to Cameroon. The fertile highlands of Rwanda and Burundi, extending into the Kigezi province of Uganda, support some of the highest densities of rural population on the continent, over 400 persons per square mile (150 per sq. km). The northern littoral of Lake Victoria, from Buganda and Busoga to the Nyanza district of Kenya, also supports a high density of rural population, as does southern Malawi and eastern South Africa below the Drakensberg from Zululand to the Ciskei. The high concentrations of urban population on the Zaire/Zambia copperbelt and the southern Transvaal mining and industrial complex also show up.

In some areas of high-density rural population severe pressure on the land is relieved by out-migration. From Rwanda, Burundi and Kigezi people move northwards, extending intensive cultivation into the Ugandan provinces of Ankole and Toro. From the overpopulated South African 'homelands' people move to the towns, some permanently, others in the migrant labour system. As African populations increase, pressure on the land becomes widespread, people are forced to move and related problems multiply, as in Kenya.

The population of Africa is growing rapidly and could almost double by the end of the century. There are exceptionally high birth-rates and high, but declining, death-rates. Infant mortality is very high because of disease and poor nutritional standards. Life expectancy is low but rising (refer to Chapter 46, Development and population, and Chapter 72, Population statistics).

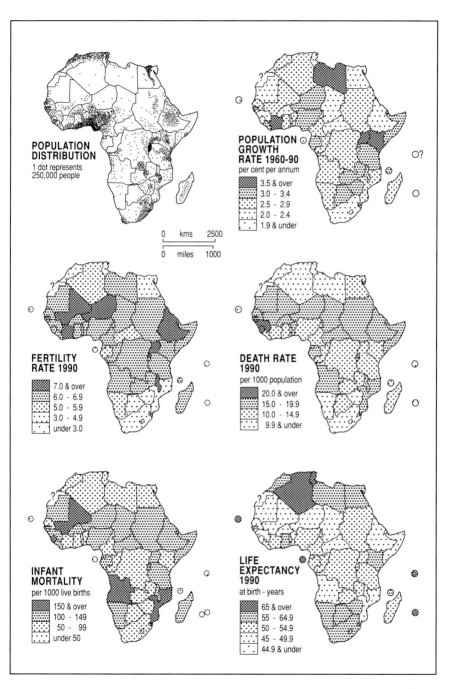

POPULATION DISTRIBUTION
1 dot represents 250,000 people

0 kms 2500
0 miles 1000

POPULATION GROWTH RATE 1960-90
per cent per annum
- 3.5 & over
- 3.0 - 3.4
- 2.5 - 2.9
- 2.0 - 2.4
- 1.9 & under

FERTILITY RATE 1990
- 7.0 & over
- 6.0 - 6.9
- 5.0 - 5.9
- 3.0 - 4.9
- under 3.0

DEATH RATE 1990
per 1000 population
- 20.0 & over
- 15.0 - 19.9
- 10.0 - 14.9
- 9.9 & under

INFANT MORTALITY
per 1000 live births
- 150 & over
- 100 - 149
- 50 - 99
- under 50

LIFE EXPECTANCY 1990
at birth - years
- 65 & over
- 55 - 64.9
- 50 - 54.9
- 45 - 49.9
- 44.9 & under

27

10 Languages

Up to two thousand different languages are today spoken in Africa, which 'displays a greater degree of overall linguistic complexity than any other continent'. The distribution of languages is complex with areas of linguistic overlap and fragmentation and, at any one place, 'layers' of language, each fulfilling a particular purpose. Africans' linguistic facility is remarkable. The barely literate house-servant at Makerere would speak to her family in RuToro, to neighbours in Luganda, to traders in Swahili, to her employers in English and, to her employers' amazement and near-monoglot embarrassment, to their visitor in fluent French: her former husband was Rwandaise!

The distribution of African languages bears no relationship to modern political boundaries and individual language groups are often divided between states. Somalia and Swaziland are among the few states to have just one African language. The people of Nigeria between them speak almost four hundred different languages. Creation of national unity without a unifying language is difficult, and most states in Africa have fallen back on the former colonial language or an Africanized form of it as the official language of the state.

Some states have a major local language as the official language. Swahili, the trading language of the east African coast, is the official language of Tanzania although it is not the hearth language of most Tanzanians. Somalia, in making Somali the official language, had to standardize its written form as recently as 1974. Other states use a local language officially alongside the former colonial language. For black South Africans Afrikaans is the language of oppression and other African languages have long been the tools of the 'divide and rule' apartheid policy. Urban blacks demand to be educated in English and in 1976 many were prepared to die for that in Soweto.

The broad distribution of African languages can be summarized as of two blocs, northern and southern, with a large fragmentation zone between. The northern bloc (which extends into the Middle East) includes the *Afro–Asiatic* languages, mainly Arabic, and also Amharic which is the official language of Ethiopia. The southern bloc embraces the *Bantu* languages, of which there are over 400, including 'families' of languages such as the Nguni in the south east containing Xhosa, Zulu and Swazi. *Nilo–Hamitic* languages, extending from northern Chad to western Kenya, are part of the fragmentation belt. The *Khoisan* languages of the San (Bushmen) and Khoi-Khoi (Hottentots), whose 'clicks' are found in southern Bantu languages which were in contact with them, are now limited to the Kalahari. *Malagasy* languages, limited to

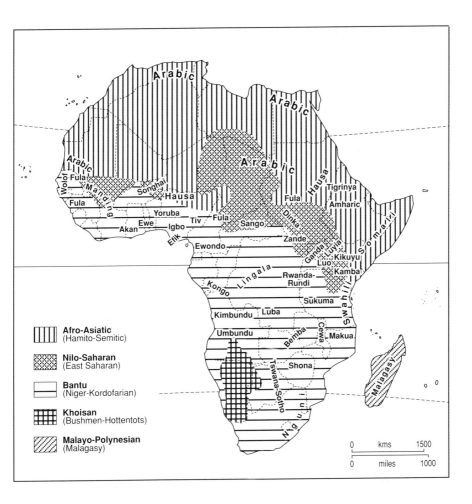

Afro-Asiatic
(Hamito-Semitic)

Nilo-Saharan
(East Saharan)

Bantu
(Niger-Kordofarian)

Khoisan
(Bushmen-Hottentots)

Malayo-Polynesian
(Malagasy)

Madagascar, are related to Indonesian, so stimulating speculation about Kon Tiki-like migrations across the Indian Ocean.

The preservation of African languages is beginning to be a problem, with increasing mobility and communication, and is becoming a cultural issue. Fifty African languages are each spoken by more than one million people and they are vital to modern Africa. Languages might be likened to plant species inasmuch as the elimination of any one could do untold and as yet unknown harm. They form a vital part of the common human cultural heritage, too valuable to be lost.

11 Literacy

In July 1966 an impressive ceremony was held in Lusaka to install the first Chancellor of the University of Zambia. During a long speech, the Chancellor, President Kaunda, publicly wept for the educational opportunities denied to his and earlier generations in colonial times and vowed that his government would right past wrongs. Many of his audience, which included academics bearing fraternal greetings from universities all over the world, were embarrassed: some because he cried, others because he really did have something to cry about.

Education is not a panacea for all the ills of Africa. There may indeed have been an over-reaction to colonial neglect, so that many in Africa have caught the 'diploma disease'. Emphasis is wrongly placed on academic rather than technical subjects and on qualifications unrelated to the jobs available. Employers, including governments, make inappropriate qualification demands of job candidates. The system raises expectations which cannot be fulfilled, is geared mainly to urban aspirations and so contributes to problems of accelerating urbanization and serious qualitative rural depopulation.

At the root of the educational problem, literacy itself ought to be a fundamental human right: not because it is a job qualification but because of its intrinsic value, its basic contribution to the quality of human life. But this is not a view shared across cultures. Of six African countries in 1990 which showed adult literacy rates of 25 per cent or less, four are largely Muslim. Here, female literacy rates are tiny (in Somalia and Burkina Faso, only 6 per cent), mainly for cultural reasons. But states with the lowest literacy rates are also amongst the poorest. In the Sahelian states and Somalia, in addition to Islam and poverty, the nomadic life practised by many contributes to educational problems. The Gambia, Guinea and Senegal are also poor and Muslim, a potent combination of factors making for low literacy rates. The richer Muslim states of north Africa fare better (Libya 64 per cent, Tunisia 65 per cent, Algeria 57 per cent literacy). Angola, Ethiopia, Liberia and Somalia sacrifice literacy to war and poverty.

Although slowly improving, low literacy rates are common through Africa. Twenty-seven states have adult literacy rates above 50 per cent, with Lesotho, Madagascar, Mauritius and Seychelles over 70 per cent. These figures correlate well with statistics of newspaper circulation, to revealing interesting traditions arising from and encouraging literacy.

In 1986 Uganda and Ethiopia committed the least proportion of their gross national product (GNP) to education at 1.5 per cent and 2 per cent

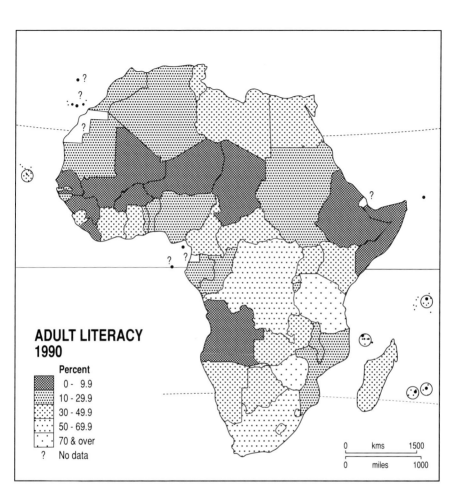

ADULT LITERACY 1990

Percent
- 0 - 9.9
- 10 - 29.9
- 30 - 49.9
- 50 - 69.9
- 70 & over
- ? No data

0 kms 1500

0 miles 1000

respectively. In contrast thirteen African states spent more than 5 per cent of GNP on education. Political upheaval and civil war are clearly not conducive to investment in education. However, African states are not alone in giving a lower priority to education than to guns.

Literacy is the key to communication; it is only partly substituted by oral means such as radio and public meetings. Only through literacy can people acquire the skills necessary to run mines, railways and factories and to improve agriculture. How many (especially women) in rural Gambia can read notices proclaiming: 'Islam is not against family planning'?

12 Quality of life

In 1990 the United Nations Development Programme (UNDP) published its first *Human Development Report*, which broke new ground by attempting to measure the quality of life enjoyed by people throughout the world rather than merely the economic performance of the individual states. Instead of using only the usual economic indicators, such as GNP per caput, a Human Development Index (HDI) was calculated for each state; this attempted to answer the question, 'how are the people faring?' rather than simply, 'how much is this nation producing?' The 1991 Report refined and extended the concept to include measurement of gender disparities, income distribution, human progress and human freedom, but retained the Central Index based on three fundamental components, 'longevity, knowledge and decent living standards', as expressed through figures for life expectancy, literacy and income.

The basic facts about global distribution of income are shocking: 77 per cent of the world's population earns 15 per cent of the world's income. Within that there is a 'steady concentration of poverty in Africa'. While economic growth in sub-Saharan Africa has been very sluggish, population growth is high at an average of over 3 per cent per annum. The net result is that per caput income has fallen by over 2 per cent per annum. Nevertheless, important gains have been made in terms of human development. Since 1960 life expectancy in sub-Saharan Africa has increased from 40 to 52 years; literacy rates have increased by two-thirds between 1970 and 1985. But, according to the Report, 'the outlook for Africa is bleak unless concerted national and international efforts set the continent on a more positive course'. Impediments to human development include: apartheid, 'cross-border conflicts, ethnic upheavals and civil strife'. All of these result in six million refugees and 50 million disabled people, whilst a further 35 million people are displaced through natural disasters and 'difficult economic conditions'. Improving the position of women is one of the most important tasks facing Africa, where female literacy rates are 34 per cent compared with 56 per cent for males and primary school enrolment is 44 per cent compared with 54 per cent for males.

On the world scale, African countries fared poorly: in 1992 none made the HDI 'High' category; just eight, Mauritius, South Africa, the Seychelles, Libya, Tunisia, Botswana, Gabon and Algeria, were placed in the 'Medium' range; the remaining 45 were in the 'Low' category. Eighteen of the world's bottom twenty were from Africa, with an HDI of 0.166 or less, on a scale where under 0.500 is classified as low and the highest state is at 0.982. Of the

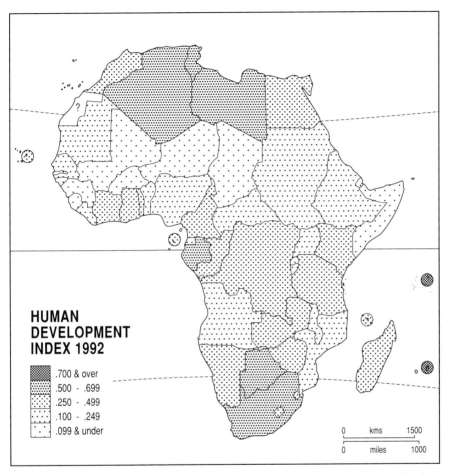

HUMAN DEVELOPMENT INDEX 1992

- .700 & over
- .500 - .699
- .250 - .499
- .100 - .249
- .099 & under

| 0 | kms | 1500 |
| 0 | miles | 1000 |

eighteen worst-off African states all but two had a life expectancy at birth of less than 50 years, all eleven had an adult literacy rate of under 30 per cent and all had low incomes.

In many African states HDI performance ranked far worse than GNP performance. Libya, Gabon, Algeria, Cameroon, Sudan, Angola, Djibouti, Mauritania and Guinea were prominent in this respect, pointing to the fact that their attainments in terms of life expectancy and literacy did not match their economic achievements. The relationship between wealth and human development is not simple because factors such as war, natural disaster, personal freedom, environmental degradation and different cultural values intervene and must be accounted for in any attempt to assess the quality of life (refer to Chapter 71, HDI statistics).

13 Further reading

Barnett, T. and Blaikie, P. (1992) *Aids in Africa: Its Present and Future Impact*, London: Belhaven.

Buckle, C. (1978) *Landforms in Africa*, London: Longman.

Carpenter, R. (1973) *Beyond the Pillars of Hercules*, London: Tandem. (First published: New York, Delarcorte, 1966.)

Chirimuuta, R.C. and Chiramuuta, R.J. (1987) *Aids, Africa and Racisim*, Bretby, Derbyshire: privately published.

Dalby, D. (1977) *A Language Map of Africa*, London: Royal Africa Institute.

Glantz, M.H. (ed.), (1987) *Drought and Hunger in Africa: Denying Famine a Future*, Cambridge: Cambridge University Press.

Griffiths, J.F. (ed.) (1972) *Climates of Africa*, London: Elsevier.

Grove, A.T. (1986) 'The state of Africa in the 1980s', *Geographical Journal* 152(2): 193–203.

King, L.C. (1951) *South African Scenery*, Edinburgh: Oliver & Boyd.

Pocock, G.N. (ed.) (1932) *Herodotus: Stories and Travels*, London: Dent.

Ransford, O. (1983) *'Bid the Sickness Cease': Disease in the History of Black Africa*, London: Murray.

United Nations Development Programme (UNDP) (annual from 1990) *Human Development Report*, New York and Oxford: Oxford University Press for UNDP.

Wellington, J.H. (1955) *Southern Africa: A Geographical Study*, vol. 1, Cambridge: Cambridge University Press

B Historical

14 Africa: cradle of humankind

Archaeologists working in Africa are literally unearthing and piecing together a significant refinement of the theory of human evolution. With due allowance for the ongoing nature of the research and disputes about precise dating and interpretations, there is now emerging from Africa an exciting advance in our understanding of human evolution.

Charles Darwin pointed to Africa as the one place most likely to produce evidence needed to close the gap between the widely known *Ramapithecus* of about 12 million years ago and the direct ancestor of humans, *Homo erectus*, of 1 million years ago: the search was for the 'missing link'. In 1924, in a limestone quarry at Taung in the northern Cape Province of South Africa, a hominid skull was identified and named as *Australopithecus africanus*, a biped ape-man. Although reported in the scientific press (*Nature*), the discovery was largely ignored. Then in the late 1930s, in the Transvaal, other skulls also identified as Australopithecenes were found in two distinct forms, *A. africanus* at Sterkfontein and Makapansgat and the larger, more thickset *A. robustus* at Kromdraai and Swartkrans. Unfortunately, since they were found in cave sites, they were difficult to date but, apparently, here in Africa was the missing link. The hypothesis now emerging is somewhat more complex.

Since 1959 an enormous number of hominid fossil remains have been unearthed in the Rift Valley of east Africa. They include examples of *A. africanus* and *A. boisei*, the east African form of *A. robustus*, dated between 3 and 1.5 million years ago, but also examples of early *Homo* species dating to 2 million years ago. The existence of two types of *Australopithecus* led to a suggestion that the *robustus/boisei* form might be a side-shoot on the evolutionary tree. Now there is evidence that a form of *Homo* lived contemporaneously with Australopithecenes. It is hypothesized that about 5 million years ago *Ramapithecus* divided into the two *Australopithecus* forms, which later became extinct, and the *Homo* branch of *Homo habilis*, *Homo erectus*, *Homo sapiens* and modern humans, *Homo sapiens sapiens*. Some dating doubts remain and further evidence needs to be examined before the hypothesis is fully accepted.

It poses a fascinating geographical conundrum. Why, when *Ramapithecus* and *Homo erectus* are found in Africa, Asia and Europe, is *Australopithecus* found only in Africa south of the Sahara? Are remains merely awaiting discovery, perhaps even in a Sussex gravel pit, or were humans somehow confined to Africa in this period of evolution?

Homo erectus sites have been discovered throughout continental Africa

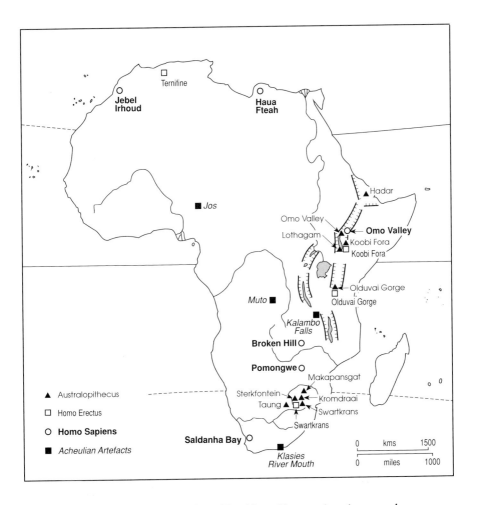

including the Maghreb, notably at Ternifine. *Homo sapiens* sites are also very widely distributed in Africa, from Saldanha Bay to Haua Fteah. Another clue to human pre-history are tools or artefacts. Study of Acheulian artefact sites shows that they were very widespread in Africa and that the use of these tools spread from Africa to Asia and Europe where they represent the first local tool-making tradition. Is not this further evidence pointing towards Africa as the 'cradle of humankind'? Elsewhere, notably in China and Pakistan, despite considerable archaeological effort using modern excavation and dating technology, nothing indisputable has yet emerged in the key form of hominid remains seriously to challenge the Africanist assertion.

37

15 Pre-European history

As Africa was a cradle of humankind so Egypt was a cradle of civilization. By using simple agricultural techniques, primitive metal tools and the annual flood-waters of the Nile, the Egyptians created one of the earliest, richest and most durable civilizations. Flood irrigation was the key to the Egyptians' ability to produce a regular surplus of food. Their society and polity grew in complexity and flourished for thousands of years, from before 3000 BC to the Assyrian conquest of 665 BC. The achievements of early Egypt are remarkable, for example in writing, medicine and architecture, with all of the skills of the arts, mathematics, science and engineering implied. The geographical core of Egypt was the Nile delta and lower valley. At its greatest extent the Egyptian empire stretched from Nubia to Syria.

The Assyrians were the first of a succession of Eurasian empires to conquer Egypt and other parts of north Africa. The Persians led by Cambyses came in 525 BC. Under Darius they completed the first Suez canal, from the Gulf of Suez to the Nile, and established a regular sea-trading route from Egypt to Persia via the Erythrean (Red) Sea.

The Persians sponsored the Phoenician circumnavigation of Africa that was recorded by Herodotus. The Phoenicians, fabled seamen and sea-traders, sailed throughout the Mediterranean, exploited silver mines in Spain, established trading settlements on the Atlantic littoral and founded an empire based on Carthage which eventually became independent.

The Greeks also set up trading settlements on the north African coast as at Cyrene. Alexander the Great founded Alexandria (332 BC) as a new provincial capital. It stands as one of the first examples of the colonial port-capital city, so familiar in modern Africa. On Alexander's death Egypt became the centre of the kingdom of Ptolemy, one of Alexander's generals. Greek settlements were established as far down the Erythrean coast as Adulis.

The Roman empire was the first to include the whole of the north African coast and inland as far as the desert. The Romans had to overcome the Carthaginians led by Hannibal. He was defeated at Zama south of Carthage in 202 BC. In 30 BC they finally took Egypt. Rome held sway over north Africa for about 400 years until Rome itself fell. Even then Egypt remained part of the east Roman empire ruled from Constantinople. It was a Christian empire whose missionaries not only made Egypt into a Christian kingdom but also converted the people of Nubia.

Within ten years of the death of the Prophet Mohammed the Christian kingdom of Egypt fell to the Arab Caliphate, inspired by the new religion of

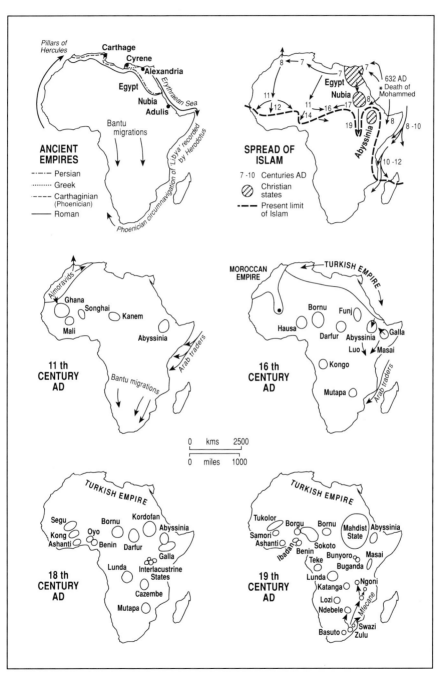

ANCIENT EMPIRES

Pillars of Hercules
Carthage
Cyrene
Alexandria
Egypt
Nubia
Adulis
Erythraean Sea
Bantu migrations
'Libya' recorded by Herodotus
Phoenician circumnavigation of

-·-·- Persian
········· Greek
- - - - Carthaginian (Phoenician)
——— Roman

SPREAD OF ISLAM

8   7   7   7
Egypt
Nubia   632 AD
* Death of Mohammed
11   8
12   11   16   17   8
14   19   8   8-10
Abyssinia
10-12

7-10   Centuries AD
⊘   Christian states
- ▪- ▪-   Present limit of Islam

11th CENTURY AD

Almoravids
Ghana
Songhai
Mali   Kanem
Abyssinia
Arab traders
Bantu migrations

16th CENTURY AD

MOROCCAN EMPIRE
TURKISH EMPIRE
Bornu   Funj
Hausa
Darfur   Abyssinia   Galla
Luo   Masai
Kongo
Mutapa
Arab traders

0   kms   2500
0   miles   1000

18th CENTURY AD

TURKISH EMPIRE
Segu   Kordofan
Bornu
Kong   Oyo   Abyssinia
Ashanti   Benin   Darfur
Galla
Lunda   Interlacustrine States
Cazembe
Mutapa

19th CENTURY AD

TURKISH EMPIRE
Tukolor   Borgu   Bornu   Mahdist State   Abyssinia
Samori
Ashanti   Sokoto
Ibadan   Benin   Bunyoro   Masai
Teke   Buganda
Lunda
Katanga   Ngoni
Lozi   Mfecane
Ndebele
Basuto   Swazi
Zulu

39

Islam. By AD 705 the Arabs had conquered as far west as Morocco. Also in the eighth century Islam began to spread along the sea-trade route from the Horn of Africa down the east African coast. The Nubian Christian kingdom and Christian Abyssinia delayed the westward spread of Islam but the Sahara was not as great a barrier. By the eleventh century Islam had crossed the desert via trade routes and then spread east and west along the Sahelian corridor between desert and forest. Nubia and the upper Nile valley was overwhelmed much later to complete the spread of Islam in Africa to its present limits.

Meanwhile, in Africa south of the sahara, large-scale, long-distance migrations of peoples were taking place. The Bantu-speaking peoples moved eastwards and southwards from west Africa. By the eleventh century they reached southern Africa where they came into contact with the San (Bushmen) and Khoi-Khoi (Hottentots) who were gradually pressed into the south-western corner of the continent.

Along the Sahel several African states arose. Among the earliest in the west were Ghana and Mali, names taken by modern states, with some geographical licence, to acknowledge their political and cultural roots extending back over 1000 years. In the eleventh century the western Sahelian states were over-whelmed by the Almoravids, a puritanical Islamic sect based in Mauritania, who vigorously pursued their cause throughout north-western Africa. At the southern end of well established trans-Saharan trade routes, the Sahelian states often had to endure such incursions from the north; it was a locational risk. Trade flourished in gold, salt, ivory and slaves while the grasslands of the Sahel and the fertile area of the Niger inland delta were well able to support the urban populations of the trading cities.

The precise role of European sea-trading posts, established along the west African coast from the late fifteenth century, in the decline of the Sahelian trading towns is disputed, but as products of the west African hinterland were attracted to the coast by the higher prices offered by Europeans so the Sahelian towns found themselves on the periphery of trade not at the centre of it as they had been. As time has passed the effect of that peripheral location has become more marked and the prosperity of trading towns, such as Timbuctoo, has dwindled.

In 1591 the Songhai empire was defeated by any army from Morocco which had crossed the Sahara. The western Sahel became Moroccan and although the link with Morocco became more tenuous over the centuries the Moroccan hegemony of that time forms the basis of modern Moroccan claims to the western Sahara, Mauritania and parts of Algeria.

The Sahelian belt spawned many new states including Hausa-land and Bornu. Abyssinia became a beleaguered island of Christianity in an Islamic sea. In north Africa the Turkish empire extended from Morocco to Eritrea.

Various Nilotic groups migrated into east Africa. The Bantu state of Kongo emerged as a powerful force in northern Angola. In the south the Mutapa of Great Zimbabwe fame not only built wonderful and mysterious stone structures but also worked copper and gold to sell with ivory to Arab traders on the Swahili coast.

In the eighteenth and nineteenth centuries small but significant states emerged in the forest fringe area of west Africa: the Ashanti, Oyo, Ibadan and Benin are important examples. They were rich; their people lived in towns and developed craft industries, including metal-working, which today form part of the cultural heritage. Ashanti gold-weights and Benin bronzes are magnificent examples of such traditions. In the Sahel the power of individual states waxed and waned. Between the great lakes of east Africa the Bantu kingdoms of Buganda, Bunyoro, Busoga, Ankole and Toro arose and in central Africa the Lunda, Katanga and Cazembe gained prominence.

In southern Africa in the early nineteenth century population pressure on the land helped to cause the emergence of several nations, including the Zulu, Swazi, Basuto and Ndebele, and a tumult of mass migrations known as the *mfecane*. At the epicentre was the Zulu kingdom built up in a remarkably short period by Shaka. In 1816, having seized the chieftainship of his own small clan, the Zulu, Shaka proceeded to fashion them into a formidable fighting force. By 1820, inheritance, conquest and absorption had made them the foremost military force in southern Africa. The *mfecane* spread in a series of shock waves from Zululand. Mzilikazi, one of Shaka's *indunas*, formed the Ndebele nation as he fled Shaka's wrath. On the high veld the Ndebele encountered the trek-boers and so migrated north to settle round Bulawayo, named after the Zulu royal kraal. Today Joshua Nkomo is Mzilikazi's political heir and Mangosuthu Buthelezi is Shaka's.

The pre-European history of Africa is fascinating, long and complex. It owes much to oral traditions which are being converted into written records. The African past *does* exist and a proper understanding of it is very often the key to current political issues.

16 European penetration to 1880

Europe's first toe-hold on the African continent was Ceuta on the north Moroccan coast, captured in August 1415 by the Portuguese. Passing to Spain with the union of the Portuguese and Spanish crowns in 1640, Ceuta remains Spanish to the present day and is, with Melilla (first occupied by Spain in 1497), Europe's last grasp on the African continent.

The taking of Ceuta was the prelude to the Portuguese voyages of discovery which opened up the sea route to India as an alternative to the route through the eastern Mediterranean that was dominated by the Venetians and Saracens. The Portuguese were motivated by a complex mix of Christian zeal against the infidel Muslims, the hope of a strategic Christian link-up with the legendary Prester John, the lure of Guinea gold, and the ultimate prize of the rich India trade. They swept down the inhospitable north-west coast of Africa – Cape Bojador (1434), Cape Blanco (1441), Cape Verde (1444) – to black Africa. By 1460 trade in gold and slaves was established and the Portuguese had reached Sierra Leone. They set up fortified coastal trading posts from Senegambia to the Gold Coast and 'captured' much of the ancient trans-Saharan trade. Diogo Cão 'discovered' the mouth of the River Zaire (Congo) in 1484 and in 1486 reached Cape Cross. Bartholomeu Dias rounded the Cape of Good Hope and planted his furthest *padrao* (stone cross) at Kwaaihoek in March 1488. After a mysterious nine-year-delay, Vasco da Gama rounded the Cape, sailed past Natal on Christmas Day 1497 and arrived at Malindi in April 1498. Picking up a Gujerati pilot, da Gama had a quick passage to India, arriving near Calicut on 20 May 1498.

On the east African coast the Portuguese took over a string of Swahili trading posts from Sofala and Quelimane to Mombasa, Malindi and Lamu. The main trade of the east coast was gold and ivory but in rather disappointing quantities, which discouraged penetration of the interior except along the lower Zambesi valley. The Portuguese did establish contact with the Prester John, leader of Christian Ethiopia, but he was far less powerful than they had hoped and they had to help him against the forces of Islam which threatened on all sides.

On the west coast trade flourished in gold, silver, ivory, hides and, above all, slaves. From their coastal trading posts and two great slave entrepôts of São Tomé and Santiago the Portuguese shipped slaves to their plantations in Brazil and to Spanish America. Later Luanda (1575) and Benguela (1617) also became prominent in the slave trade, and from these the Portuguese penetrated the hinterland of Angola.

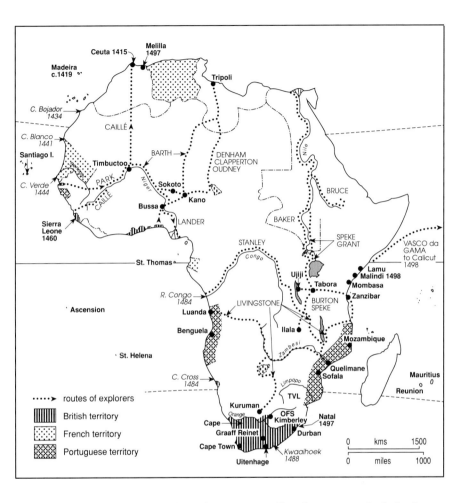

The Portuguese metropolitan base was small and over-stretched. In the seventeenth century their supremacy was challenged as East India companies were formed by the British (1600), Dutch (1602), French (1604) and Danes (1610). The Dutch actually held Luanda and Benguela between 1641 and 1648 before establishing their own victualling station at Cape Town in 1652. The British set up a station at St Helena (1659) after the Dutch; the French at Ile de Bourbon (Réunion) in 1642 and in 1715, also in succession to the Dutch, at Ile de France (Mauritius). All of the European maritime nations established trading posts on the west African coast but their interest was limited to the coast and to navigable rivers such as the Gambia. Up-country trading was generally conducted by proxy. There was little further penetration of the interior except by the French in Senegal and by the Dutch from the Cape.

Within six years of the Dutch East India Company setting up its Cape Town station there were free burghers at the Cape. By the end of the eighteenth century Dutch farmers (*boers*) had spread slowly eastwards into the unpromising interior, practising extensive cattle-ranching in a semi-arid environment to establish small settlements at Graaff Reinet (1786) and Uitenhage (1804). To keep the Cape out of French hands in the Napoleonic wars the British took it from the Dutch (1795), returned it to the Batavian Republic (1803), and finally retook it in 1806. In 1814 Britain kept the Cape, paying the Dutch an indemnity of £6 million. It was a fateful decision which, over a generation, led to the opening up of the interior by the trek-boers. Disgruntled with British justice, with British incompetence and inconsistency on the frontier, with the abolition of the slave trade (1807) and then slavery itself (1833) and with British trickery in paying what was felt to be inadequate compensation in London, some Boers decided to trek away form British rule in the Cape Colony (1836). They crossed the Orange river, eventually to found the Orange Free State (1858). Some went on across the Vaal and later formed the South African (Transvaal) Republic (1854), while others crossed the Drakensberg down into Natal to found the Republic of Natalia (1841). Within months the European frontier in southern Africa was extended inland by a thousand miles, if a little precariously. Many Africans were displaced in long-drawn-out conflict which involved the strongest groups, the Zulu, Ndebele and Basuto. The British took the comparatively easy sea route to Durban, claimed Natal for their own (1843) and pushed out the Boers. When diamonds were found on Boer farms east of the Vaal river in 1870, Griqualand West (Kimberley) quickly came under British 'protection': capitalism and imperialism were beginning to walk hand in hand.

The London Missionary Society had preceded the trek-boers over the Orange river to found their most famous mission at Kuruman (1816), from which the next stage of European penetration of southern Africa began in 1841 with David Livingstone. First to Lake Ngami then to the Zambesi at Sesheke and on to Luanda on the west coast Livingstone tramped: he retraced his footsteps to Sesheke, then travelled down the Zambesi to 'scenes such as angels must have gazed upon in their flights', and named the Victoria Falls (1855). He continued down the great river to Quelimane where the inevitable British brig soon came to return him to London. Livingstone was the first European to cross Africa but was not the first European, or even Scots, major explorer of Africa.

James Bruce had followed the Blue Nile into the highlands of Ethiopia in 1766 and in the west Mungo Park had reached the upper Niger from the Gambia in 1796 and returned to sail down that river to his death at Bussa in

1806. Denham, Clapperton and Oudney crossed the Sahara from Tripoli to Kano and Sokoto; Clapperton with Lander returned to Kano via the west African coast only to die there; Lander with his brother sailed down the Niger from Bussa to its mouth: all in the period 1823–30. In 1827 René Caillé set out, in disguise, for the fabled golden city of Timbuctoo but the reality he reported was so dull and earthy that few believed he had even been there. Heinrich Barth, the German emissary of the Royal Geographical Society, finally laid the legend of Timbuctoo in a five-year expedition reported in five stout volumes.

Meanwhile the quest for the source of the Nile was under way: Burton and Speke from Zanzibar to Lake Tanganyika, Speke to Lake Victoria, Speke and Grant down the Nile, to breakfast with Sam Baker and his lady on their grand way to Lake Albert, and then home to dispute with Burton and the armchair geographers. Livingstone returned to the Zambesi but it proved not to be 'God's highway to the heart of Africa' as he had hoped, so he turned northwards to ensure that Malawi would be a second home for future Scots missionaries. On his last journey Livingstone continued his crusade against the Arab slave trade and joined unsuccessfully in the Nile quest; he died, a lonely, broken man, at Ilala in 1873.

Livingstone himself had been 'discovered' at Ujiji by Stanley in 1872, with the incredible greeting, 'Dr Livingstone, I presume?' Stanley, fatherless son of a Welsh workhouse, 'adopted' Livingstone and his theory that the Lualaba was in fact the Nile. He got off to a bad start when the then president of the Royal Geographical Society declined even to believe that he had found Livingstone. Thanks in part to the intervention of Queen Victoria herself, Stanley survived to become the greatest, if one of the most ruthless, of all of the European explorers of Africa. He circumnavigated Lake Victoria and then Lake Tanganyika, crossed to the Lualaba and followed it 2000 miles (3200 km) 'through the dark continent' and after 999 days reached the west coast. Alas, Livingstone had been wrong, but the mystery of the Nile and the Congo had finally been resolved in one epic, brutal journey. Stanley twice returned to Africa on major expeditions which were both closely associated with the next phase of Europe's affair with Africa – the scramble. The explorers collectively combined missionary zeal with scientific enquiry and were not averse to keeping an eye open for prospects of trade. They excited European interest in Africa with their fascinating tales and wonderful books and paved the way for the impending partition of Africa between the European powers.

17 Slave trades

After braving the hostile, barren, western Sahara coast, the Portuguese in the mid-fifteenth century reached in west Africa a green land (Cape Verde), rich in prized items of trade (Ivory Coast, Gold Coast, *Slave* coast). West African slaves were sold in Lisbon as early as 1444 but it was not until the Americas were opened up in the sixteenth century that the Atlantic slave trade flourished. This infamous trade prospered because in the Americas plantation agriculture, producing cash crops such as coffee, cotton, sugar and tobacco for the European market, created an almost insatiable demand for slaves. The supply of slaves in west Africa presented no great problem, so that European slave-traders could, and did, make enormous profits.

In its most notorious and highly developed form the slave trade was a 'triangular' trade. Cheap manufactures from Europe, mainly England, were traded on the west African coast with African middlemen for slaves, who were cargo for the 'middle passage' to the Americas where they were sold. With the proceeds were purchased products of slave labour, sugar, cotton and tobacco, to sell in Europe at great profit. A small reinvestment in a cargo of manufactures started the cycle all over again. While the English were the most prominent, all European nations indulged in the slave trade: Danes, Dutch, French, Germans, Portuguese and Spaniards. The Portuguese established slave entrepôts at São Tomé and Santiago but their greatest trade was from Angola to Brazil; Luanda was for a long time the largest slave port on the west African coast. Early in the nineteenth century as many as 135,000 slaves a year were taken across the Atlantic.

The depredations of the slave trade make a catalogue of horror, from the basic human indignity to the brutal decimation of whole populations, notably that of Angola. Village was set against village, tribe against tribe in violent campaigns of terror fought for human booty. Near the end of the eighteenth century, revulsion towards it all began to spread among the more enlightened leaders in Britain. A long hard struggle, made a little easier by economic doubts about the trade itself, gained first the abolition of the slave trade (1807) and then the abolition of slavery itself (1833). In the tradition of setting a thief to catch a thief, the ubiquitous British navy ('African squadron') kept itself in fighting trim through the middle years of the nineteenth century by stamping out the slave trade wherever it could.

That included the Arab slave trade on the east coast of Africa, centred for centuries on Zanzibar. The Arab traders, unlike their European counterparts in the West, penetrating deeply into the hinterland. European explorers

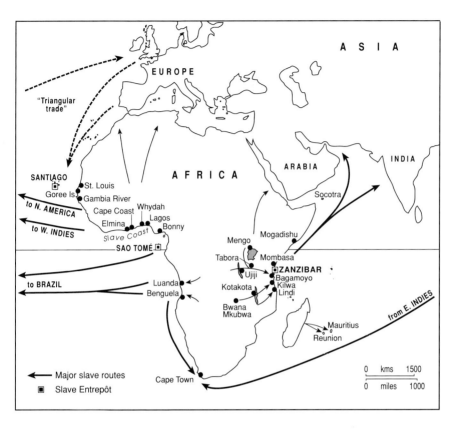

encountered Arab slave parties in Malawi and Zambia, along the well-defined
Bagamoyo to Ujiji trade route, in Uganda and the Sudan and as far west as the
Lualaba river, where the slave trader Tippu Tib held sway to the discomfiture
of Stanley's ill-rated 'rear column' as late as 1888.

The significance of four centuries of slavery is enormous: its contribution
to European economic development and African underdevelopment, to racial
prejudice and to the creation of multiracial societies. Slavery persisted into
this century, indeed in some forms it survives today; thus the Anti-Slavery
Society of London remains active.

18 The scramble for Africa

There was a marked geographical symmetry to European colonization in Africa as settlers were first attracted to the warm temperate climates of Algeria (1830) and South Africa (1652). Then in 1869 the opening of the Suez canal and the diamond rush in South Africa simultaneously transformed both ends of the continent.

The significance of diamonds to the subsequent history of southern Africa and to the scramble is not easily exaggerated. Britain had taken the Cape during the Napoleonic wars to safeguard the route to India. It was inconvenient that the Cape station was part of a colony, and costly and painful that the colony should have an unstable frontier disputed by expansionist Boers and expansionist Xhosas. British imperial attitudes waxed and waned. Money was reluctantly voted, frontier wars were waged and territory annexed; but money was also withheld, peace made and land returned. Such changes often reflected changes in government at Westminster and swung with the fortunes of Tories and Liberals. The most dramatic swing came in 1881 when, after a period of expansion which included the annexation of the Transvaal (1877) and the Zulu war (1879), the newly elected Gladstone government made peace following the British defeat by the Boers at Majuba, granting the Transvaal its independence but retaining a not easily defined British 'suzerainty'. The net result of the swing of the imperial pendulum in South Africa was, however, of British territorial expansion on the basis of two steps forward, one step back.

The diamonds of Kimberley added a new dimension to British imperialism in Africa, a financial rationale for territorial expansion which united local settler interests with metropolitan capitalists. It facilitated the subsequent exploitation of the Witwatersrand goldfields (1886) and, through Lenin's adaptation of Hobson's polemic, gave rise to the thesis of 'imperialism the highest form of capitalism'. The botched Jameson Raid (1895) and the Anglo–Boer War (1899–1902) were classical manifestations of capitalist imperialism. The irony is that the immediate victims were the Boers who had been cheated out of the diamond-fields by an outrageous boundary 'adjustment', then steam-rollered by a vast imperial army dedicated to the cause of capitalism cloaked in the most appalling jingoism. The Boers had already carved their settler republics out of conquered lands and they were now quick to learn from the British.

The Suez canal was the key to the Middle East route to India. A personal triumph for Ferdinand de Lesseps, the canal was engineered in the face of determined opposition from Britain, which feared the long-term strategic

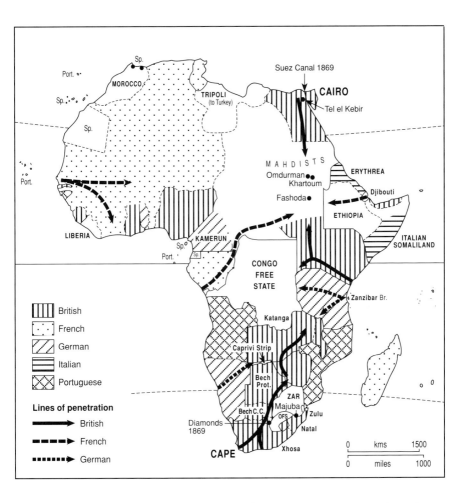

consequences of French control. Disraeli's swift move of 1875 to buy the Khedive's shares in the Canal Company for £4 million was, in fact, welcomed by de Lesseps and in 1876 Britain and France assumed joint control of Egypt's financial affairs. In 1882 Britain, to safeguard her investment, intervened against the Arabi revolt, won the battle of Tel-el-Kebir and occupied Egypt. France, having declined to join in military intervention, 'faded politically from the Egyptian scene'.

Egypt had conquered the Sudan in 1820–1 and had built-up regionally based administrations which, in later years, employed British and other European officers. In 1881 the revolt of the Mahdi began and spread throughout the Sudan, leading to the expulsion of the Egyptians and the death of General Gordon at Khartoum (1885). Emin Pasha, Governor of Equatoria Province, became a refugee from the Mahdi's forces and was unwillingly 'rescued' by the intrepid Stanley in 1889 on his last major expedition 'in darkest Africa'.

Britain, now in occupation of Egypt, eventually determined to push up the Nile. Security of the route to India, for 300 years a main plank of British foreign policy, now required control of Suez, which in crisis required the occupation of Egypt, which in turn was secure only when the Sudan was made safe, given that Uganda, at the source of the Nile, was already 'protected' (1894).

The final act of the dispute between Britain and France on the Nile was near comic opera. In 1896 the French sent a small party under Major Marchand on an epic trans-continental journey from the Congo to the Nile, there to meet with a second French force moving west from Djibouti with Ethiopian support. Against all odds Marchand reached Fashoda on the Nile in 1898 to await the Djibouti force. It failed to arrive, having been decimated by disease and sickness on descending to the hot plains from the Ethiopian highlands. Instead, up the Nile, fresh from his triumph over the Mahdi at Omdurman (September 1898), came Kitchener at the head of a large Anglo-Egyptian army. The French force of eight officers, three NCOs and 130 Senegalese troops stood its ground to act out the diplomatic charade of the 'Fashoda incident'. Eventually the French withdrew with honour, leaving the British to control the Nile from Uganda to the sea, a position confirmed by the French in the 1904 *entente* in return for a 'free hand in Morocco'.

Rivalry between the European powers for reasons of trade, strategic advantage or simply national prestige was at the heart of the scramble for Africa. Bismarck's extension of German imperial protection to the Bremen trader Lüderitz in South West Africa in 1883 precipitated a flurry of British activity of both imperial and local South African varieties. The perceived danger was of Germany from the West linking with the Boer republics to cut 'the road to the north'. In quick succession the western boundary of the Transvaal was fixed, the two tiny Boer republics of Stellaland and Goshen were eliminated, Bechuanaland Protectorate was proclaimed and the Crown Colony of Bechunaland was annexed by the Cape Colony.

The discovery of the immensely rich Witwatersrand goldfields in 1886 further whetted British appetites for the drive into the interior. In 1890 Rhodes, armed with a Royal Charter and dubious 'concessions', launched his pioneer column to Mashonaland. The Ndebele were conquered in 1893 and again in 1896 with the aid of the Maxim gun. Amid cries of 'Cape to Cairo', the railway reached Bulawayo in late 1897 and British South Africa Company rule was extended across the Zambesi.

Meanwhile, in 1884–5 the European powers met in Berlin to construct the rules for the 'great game of scramble'. They declared a free trade zone across Africa, encompassing entirely the Congo Free State which became in effect the private colony of Leopold, King of the Belgians, through his International Association. 'The General Act of the Conference of Berlin' recognized European 'spheres of influence' in Africa and provided such rules as 'occupations on the coast of Africa in order to be valid must be effective, and any new occupation on the coast must be formally notified to the Signatory Powers.'

The particular motives for declaring a sphere of influence varied as Europeans, who were convinced of their 'civilizing role', the 'truth' of their religion and their right to trade, strove to exploit Africa and to compete with each other. For the most part this proved harmless for themselves but not for Africa. For the Europeans it *did* become a gigantic game, some super 'Monopoly', played with real land and real people. Zanzibar was traded for Heligoland, parts of Northern Nigeria were exchanged for fishing rights off Newfoundland, Cameroun became Kamerun for a 'free hand in Morocco'.

Almost all of the European nations had established trading posts along the west African coast. By the mid-nineteenth century many had ceased trading and some rationalization of European interests was achieved with the British and the French most influential, but they failed to agree to a proposed grand division between themselves. Penetration from the coast was a slow process and came with sharp, punitive expeditions such as that led by Wolseley against the King of Ashanti at Kumasi in 1873. When spheres of influence were established under the rules of the scramble they were extended inland from short stretches of coast, giving rise to a pattern of long narrow colonies. The French advanced inland from Dakar on the Senegal–Niger route and were thus able not only to forestall British penetration of the far interior but also to link up with their own colonies in Ivory Coast and Dahomey. West Africa thus became the most fragmented part of the African coastline with twelve different colonial territories between Cape Blanco and Calabar.

So Africa was divided up and provided with its sometimes strange colonial boundaries. The Germans insisted on having access to the Zambesi and so the finger of the Caprivi strip was drawn on Europe's map of Africa. The Katanga pedicle, defined by a watershed which, when officials went to look, could hardly be found on the ground, fortuitously divided the central African copperbelt. When in doubt the straight line was employed: lines of latitude and longitude or, failing them, any old straight line. The scramble was a division of Africa by Europeans for Europeans. Its geographical importance lies in the fact that the colonial boundaries became the territorial framework for African independence.

19 Colonial Africa

The political map of colonial Africa was virtually complete by 1914 and there has been little subsequent change. Within fifty years that colonial boundary mesh would become the almost exact basis for territorial division of independent Africa then to be made fossilized by resolution of the Organization of African Unity (OAU) in July 1964.

In 1910 the four British colonies in South Africa were united as the Union of South Africa to fulfil a long-term British imperial dream. Botha and Smuts would not accept the multiracial Cape franchise in all parts of the Union and so the Liberals, bending over to be fair to the Boers after the excesses of the 1899–1902 war, acquiesced, though to be fair few foresaw apartheid and the disenfranchisement of non-whites in the Cape in the 1950s. The three protectorates and Southern Rhodesia were not part of the Union, though provision was made for their future inclusion. At the end of Company rule in Southern Rhodesia in 1923 whites there voted for 'responsible self-government' rather than join South Africa. Elsewhere, in 1914 Britain took the critical decision to unite Nigeria. Although divided into three regions in 1939, Nigeria remained united.

The First World War spilled over into Africa. British, French, Belgian and South African forces variously defeated German colonial armies except that of von Lettow Vorbeck, who emerged from the east African bush at the end of 1918 as the only undefeated German general and was fêted as such on his return to Berlin. The Allied powers shared their German spoils. Togo and Kamerun were both divided between Britain and France. France received the larger part of each and ruled them as separate territories. Britain administered her 'shares' as part of the Gold Coast (Ghana) and Nigeria respectively. German East Africa (Tanganyika) became British except for Ruanda–Urundi which, being contiguous with the now *Belgian* Congo, went to 'gallant little Belgium'. To reward Italy for being a wartime ally, Britain ceded Jubaland from Kenya to Italian Somaliland. German South West Africa (Namibia), having been conquered by Botha and Smuts in 1915, was given to South Africa whose fitness for the task was not doubted. Germany's former colonies became League of Nations' mandates, to be administered by the Allies in 'sacred trust'. In 1945 that 'trust' was transferred to the United Nations, a transfer later disputed by South Africa over Namibia. Trust status was to play a significant role in the decolonization process.

Britain gave up control of the affairs of Egypt in 1922. Fuad became king, to be succeeded by his son Farouk in 1936. Britain, however, retained military

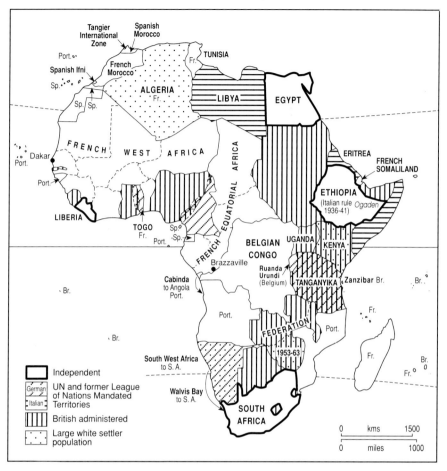

Tangier
International
Zone

Spanish
Morocco

Port. ●·

Spanish Ifni

French
Morocco

Fr.

TUNISIA

Sp.

Sp. Sp.

ALGERIA
Fr.

LIBYA

EGYPT

Dakar
Port.

FRENCH   WEST   AFRICA

NORTH AFRICA

ERITREA

FRENCH
SOMALILAND

Port.

LIBERIA

TOGO
Fr.

Sp.
Port. ·Sp.

EQUATORIAL

FRENCH

ETHIOPIA
(Italian rule Ogaden
1936-41)

BELGIAN
CONGO

UGANDA

KENYA

· Br.

Brazzaville

Cabinda
to Angola
Port.

Ruanda
Urundi
(Belgium)

TANGANYIKA ● Zanzibar Br.

Br.

Port.

Fr.

· Br.

FEDERATION

Port.

Independent

UN and former League
of Nations Mandated
Territories

British administered

Large white settler
population

German

Italian

South West Africa
to S. A.

1953-63

Fr.

Br.

Fr.

Walvis Bay
to S. A.

SOUTH
AFRICA

0   kms   1500

0   miles   1000

control of the Suez canal by means of a large garrison stationed within a newly defined Canal Zone. Ethiopia, despite having resisted Italian military conquest by winning the battle of Adowa in 1896, was by the rules of the scramble still regarded by the European powers as an Italian sphere of influence. When Mussolini came to power he was determined to reassert Italy's claim to Ethiopia. Britain and France, in the infamous Hoare–Laval pact, agreed to stand aside and in October 1935 the Italian army was launched on Ethiopia. Gallant defence was to no avail. The Emperor Haile Selassie fled to Geneva to try to stir the conscience of the League of Nations but at that time (30 June 1936) it had none.

In the Second World War Africa again became a battleground for warring European powers and again colonies changed hands. Italian forces isolated in

Somalia, Eritrea and Ethiopia were easily overcome, allowing Haile Selassie to re-enter his capital on 5 May 1941, five years to the day after the Italians had marched in. By early 1943 the Germans and Italians had been driven from north Africa. Britain assumed administrative control over Eritrea, Somalia, the Ogaden region of Ethiopia and the two Libyan provinces of Cyrenaica and Tripolitania. France administered Libya's third province, Fezzan. Ethiopia's independence was recognized but British proposals for a Greater Somalia were rejected by the United States and the Soviet Union, so Somaliland was returned to Italian administration in 1950 and the Ogaden to Ethiopia in 1955.

When the United Nations was formed in 1945 there were just four African member states: Egypt, Ethiopia, Liberia and South Africa. There were five former League of Nations' mandated territories (the former German colonies) and three other UN trust territories (the former Italian colonies). Portugal held three mainland and two island colonies; Spain two mainland and one island colony and several Moroccan enclaves; Belgium the Congo and Ruanda–Urundi. The remainder comprised the two vast colonial empires of Britain and France.

From early this century France ruled her African mainland possessions in two large federal blocks, *Afrique Occidentale Française* (AOF) 1902, and *Afrique Equatoriale Française* (AEF) 1908. Administered by governors-general at Dakar and Brazzaville respectively, the federations returned elected (African) representatives to the French National Assembly in Paris. Although they excluded the UN trust territories of Togo and Cameroun, the federations were enormous: AOF with eight territories covered 1.8 million square miles (4.7 million sq. km.), AEF with four territories covered 1 million square miles (2.5 million sq. km.), both larger than Africa's largest independent state today. The trust territories of Togo and Cameroun set the pace in political development but by the mid-1950s most of French Africa seemed very far away from independence. Then a rush of events resulted in the multiple birth of not two but fourteen new states on the African mainland, plus Madagascar. As so often happens with multiple births, several of the progeny were weak and needful of postnatal attention which France was eager to provide with intensive neo-colonial care.

Britain also experimented with federation, albeit of a very different kind, when in 1953 the Federation of Rhodesia and Nyasaland was formed between Southern Rhodesia, Northern Rhodesia and Nyasaland. Federation was strongly opposed, within the territories and within Britain itself, by many who saw it as a device for extending white settler rule. Hailed by its supporters as the new millennium the federation lasted a mere decade – a long time in African politics. The federation was seen as a bold, new Elizabethan venture in 'racial partnership' and was contrasted with the negative, apartheid state of

South Africa that the Afrikaner National Party had been grimly building since 1948. Enthusiasts pointed to the economic complementarity of the territories. The harnessing of the Zambesi at Kariba (1960), linking north and south, encapsulated the spirit of co-operation. The new (1955) direct (Malvernia) railway line to Lourenço Marques (Maputo) symbolized independence from South Africa. But critics were proved right more quickly than most had anticipated. The reality of racial discrimination mocked the ideal of racial partnership. Black political aspirations were not realized. Northern mineral resources were systematically 'ripped-off' to help to finance southern development. Federal allocation decisions almost invariably favoured the south. Arguments could always be found to support such decisions: better infrastructure, larger local market, more suitable labour, but the cumulative effect was intolerable. Federation finally founded on African nationalism, led by men who had never accepted it and saw no benefits accruing to their people or their territories. In retrospect it seems strange that the now ultra-conservative Dr Banda was the key figure in the break-up of the federation, which was the immediate prelude of independence for Malawi and Zambia.

In Southern Rhodesia colonialism, if by another name, was to continue for more than fifteen years. In all African countries with large settler populations the pattern was similar: delayed independence and violence. The French *colons* of Algeria fought to prevent independence, only to be sold out in 1962 by the autocrat whom they trusted and helped to bring to power in France, General de Gaulle. In Kenya, Mau Mau and settler resistance delayed independence until the end of 1963, two years after Tanganyika and more than a year after Uganda. This delay did not improve the prospects of the East African Community. Many common services had been developed between Kenya, Uganda and Tanganyika under British rule. Shared facilities built up over a long period included posts and telegraphs, customs and excise, railways and harbours, currency, a university and an airline. The final step of welding the territories together into a single political unit was never seriously contemplated by the British, who were eventually content to hand over a partially completed task to three states newly independent at three different dates.

Britain and France, in the administration of their African territories, saw a need to create larger economic units. Yet they gave independence to the constituent parts of their federations knowing that, once institutionalized, even the least viable of states is most unlikely to surrender its sovereignty willingly. Was it a case of deliberate 'divide and rule', as many have subsequently accused? If it was, there were many willing African accomplices each aspiring to leadership of his own small territory. On the other hand, where a federation did survive, in Nigeria, a bloody civil war had to be fought within a decade of independence in order to maintain federal unity.

20 The advance of independence

Namibia became the fifty-second African state in March 1990 and so, in little more than one generation, forty-eight new states had been born. This gives Africa more seats at the United Nations (UN) than any other continent but, despite the election in 1992 of the first African Secretary-General of the United Nations in Boutros Boutros-Ghali, very little 'clout' in international politics.

The climate conducive to decolonization in Africa evolved outside Africa. Its elements included the Second World War, the post-war exhaustion of Europe, the Atlantic Charter, the UN, the independence of India and the emergence of superpowers dedicated to ending at least overt colonialism.

The UN trust territories led the way in their respective parts of Africa. The administrative powers were accountable to the UN which created explicit, precise timetables for independence. Within the African empires of Britain, France and Belgium what was appropriate for trust territories could hardly be inappropriate for other colonies. By splendid irony the territorial booty of two world wars became the Achilles' heel of European imperialism in Africa. The Iberian empires in Africa persisted longer, held by insensitive dictatorships until their metropolitan bases crumbled.

The UN set timetables for independence in Libya (1951) and Somalia (1960), but chose for Italy's other former African colony, Eritrea, federation with Ethiopia. This led to full union in 1962, followed ever since by a secessionist war. Alone among African colonial people Eritreans were denied the right of self-determination largely because American strategic interests favoured a strong Ethiopia with a coastline. The Soviet Union supported full independence for Eritrea. In 1974 the wheel turned full circle: Ethiopia then became the client state of the Soviet Union which, in return for a Red Sea naval base, sold Ethiopia arms and military advice with which to prosecute its anti-secessionist wars. In May 1993 Eritrea achieved independence.

Egypt's revolution of 1952 hastened the British hand in the Anglo-Egyptian condominium in the Sudan which was granted independence in 1956. In Morocco, divided between France and Spain, the independence movement put pressure on the French until, increasingly embroiled in Algeria, they decided to leave. Spain followed suit but the five colonial divisions of Morocco became part of the newly independent state at different dates. Spain still retains the enclaves of Melilla and, mockingly opposite Gibraltar, Ceuta. Tunisia also gained independence in 1956 as France further cut her commitments to defend the settler colony of Algeria.

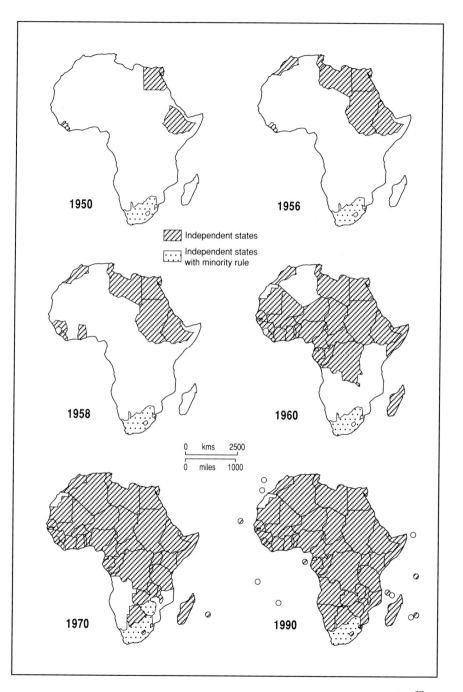

1950

1956

Independent states

Independent states with minority rule

1958

1960

0   kms   2500

0   miles   1000

1970

1990

Ghana became the first black African state to gain its independence from colonial rule in 1957. The Gold Coast had set the pace and the pattern for political development in British black Africa. Kwame Nkrumah, who was invited back from self-imposed exile to the secretaryship of the major independence party in 1947, organized demonstrations, went to jail, was released, formed his own more radical party, organized further campaigns, went to jail again, won the first election from jail, was released to head the first African administration, won further elections before internal self-government and yet more before independence when he became the first prime minister. This pattern of progress, with minor variations, was to be repeated throughout British Africa.

French Africa came to independence by a different route. The very small black colonial élite was assimilated into French culture and society and its political leaders were elected to the French National Assembly in Paris. Independence became an issue only after the *Loi Cadre* of 1956, which was passed at the initiative of the metropolitan government and applied to all colonies. The federal structures of the AOF and AEF were abandoned in favour of administration based on individual territories. This policy change was strongly supported by Felix Houphouët-Boigny of the Ivory Coast who served in the French cabinet from 1956 to 1959. Others, less conservative, denounced it as balkanization.

General de Gaulle, coming to power in 1958, gave each French African territory (except Djibouti) the stark choice of immediate, ill-prepared and unsupported independence or membership of the French Community. Guinea alone voted for independence and forthwith was ruthlessly abandoned. Two years later the other French territories, irrespective of individual preparedness, were moved *en masse* to independence. The year 1960 saw fourteen French African colonies transformed into 'independent' states. How seriously the French took this mass independence was shown by the charade enacted in the first part of August 1960 when eight French colonies successively went through flag raising ceremonies at two or three day intervals as a French presidential party hopped from one colonial capital to another.

In mid-1960 the Congo (Zaire) became independent, only to be plunged immediately into an orgy of violence. Belgium had done little to prepare the Congo and at the last moment advanced independence by one year. The few political groups to emerge were organized, divisively, on a regional basis. Within three months of independence the army seized power, arrested Prime Minister Patrice Lumumba and delivered him to secessionist Katanga (Shaba) where he was murdered in January 1961. The prolonged chaos in the Congo, the atrocities and the human suffering have to be placed at the door of Belgian politicians who, over-anxious to wash their hands of a troublesome responsibility, found themselves washing in blood.

Nigeria also came to independence in 1960 after a delay which was resolved by the adoption of a federal constitution which kept the whole colony together as a single state. It was a Nigerian rather than an imposed British solution. Nevertheless the regional difference that had so concerned the pre-independence conferences became the root causes of the civil war of 1967–70. The British empire in Africa continued to be disbanded in a piecemeal but careful fashion, each territory being treated individually according to its preparedness for independence as judged by the British: not always as objectively as claimed. In 1961 Sierra Leone became independent while the UN trust territory of Tanganyika led the way in east Africa.

The turmoil in the Congo (Zaire) was overtaken by the struggle in Algeria which had first erupted in 1954 between nationalists and white *colons*. De Gaulle, having come to power with French Algerian support, came to see the folly of attempting to hold Algeria by force of arms. He dramatically announced a cease-fire and, following the formality of a referendum, by July 1962 Algeria was independent.

Tribal differences led to the separation of Ruanda–Urundi into Rwanda and Burundi in 1962 but could not prevent post-independence massacres on a genocidal scale. In Uganda tribal differences which coincided with progressive/conservative divisions were resolved by a constitution of elaborate checks and balances which was rudely swept aside within four years of independence.

In Kenya, after a decade of sporadic violence, Britain accepted the inevitable and granted independence in 1963 under Jomo Kenyatta. Vilified, jailed, elected, released, elected again, he turned out to be a conservative leader who could even tolerate a white settler in his cabinet.

The Federation of Rhodesia and Nyasaland was broken up and, under Hastings Banda and Kenneth Kaunda respectively, Malawi and Zambia emerged. Britain also decided at this time to bring to independence a number of smaller, weaker territories of doubtful economic and political viability. The Gambia, Botswana, Lesotho, Swaziland and Mauritius all became independent in the period 1966–8. To some extent the British hand was forced because the alternative for the High Commission Territories was union with South Africa, which was nevertheless allowed to extend its economic control over them by the customs agreement of 1970.

Encouraged by British political ineptitude, white settler Rhodesia made its unilateral declaration of independence (UDI) in 1965 and for over fourteen years defied all attempts to overturn it. For all of the posturing of sanctions, with British frigates blockading Beira but ignoring Lourenço Marques (Maputo), Rhodesia's undoing was the fall of the Portuguese empire in Africa.

Once FRELIMO had achieved power in Mozambique, closed the border to Rhodesian trade and opened their territory to the Patriotic Front, Rhodesia's days were numbered. A new railway link to South Africa and a last-minute attempt at power-sharing could not save the minority regime. At the end of 1979 colonial rule was formally re-established and in April 1980 Robert Mugabe emerged as the prime minister of Zimbabwe. As with Kenyatta, and many other Africans before him, Mugabe was soon seen to be not the terrorist monster that he had been painted by sections of the British media but a leader anxious to get down to the enormous task of advancing his country and his people.

There remains the problem of the former German colony of South West Africa, Namibia. South Africa, which administered the territory, would not even recognize the passing of the League of Nations mandate to the UN and resisted independence by fighting the South West African People's Organization (SWAPO) in the field (mainly Angola), and procrastinating at the negotiating table. The issue of Namibian independence was tied inextricably to the situation in Angola. There the South Africans not only fought SWAPO but helped the *Uniao Nacional para a Independencia Total de Angola* (UNITA) against the Angolan government in Luanda who in turn were helped by Cuban troops. The Cubans and the South Africans were in effect proxies for the Soviet Union and the United States respectively. When the two superpowers came closer together one result was a 1988 agreement for both Cubans and South Africans to withdraw from Angola and for a timetable of independence for Namibia agreed under UN Resolution 345. Elections held early in 1990 led to a SWAPO victory and to independence for Namibia which had been called 'the last colony in Africa'.

Beyond Namibia is South Africa which presents a colonial problem with a difference. Independence since 1910 has been on the basis of rule by a white minority which now numbers only about 14 per cent of the population. From 1948 the Afrikaner National Party painstakingly created the apartheid state, but within forty years this experiment in blatant racial discrimination and exploitation was in ruins. The 1980s was a decade of violent repression by an increasingly desperate government trying to salvage the dream of apartheid. In contrast the 1990s has seen more realism. Under a new State President, F. W. de Klerk, in February 1990 the government recognized the African National Congress (ANC) and released its leaders, including Nelson Mandela, from life-time imprisonment. Moves towards a negotiated settlement between blacks and whites moved at a snail's pace against a background of violence largely generated by right-wing whites and their allies, the Inkhata movement of Zulu chief Buthelezi. President de Klerk, after seemingly losing ground among whites to the right wing, won overwhelming (2–1) backing in a whites-only-referendum in March 1992 to continue negotiations towards majority rule. Hopes of the process accelerating with the referendum out of the way were misplaced and, with the massacres of Boipateng in June and Bisho in September, violence seemed to be taking over. However, towards the end of September 1992 negotiations recommenced, driven in part by the time constraints on the South African government, wanting both to avoid a whites-only general election due in 1994 and to complete negotiations with an ageing Nelson Mandela rather than younger, more radical elements, and in part by the ANC leadership wishing to stay in control over those same radicals.

By mid-1993 a date of 27 April 1994, seemed likely to be confirmed for the first one-person-one-vote election in South Africa. Minds may have been concentrated by increasing violence from the white extreme right-wing with the assassination of Chris Hani and the armed invasion of the CODESA conference hall. The glaring incompetence of the South African police, found wanting at both incidents, seemed increasingly contrived. De Klerk was either unable to control sections of the police or covertly connived at police action or inaction.

All over Africa, politicians, dictators and military rulers are reluctantly willingly to relinquish power through democracy. In South Africa, de Klerk has spun out negotiations for over three years; in Malawi, Banda takes little heed of an overwhelming referendum result in favour of multi-party democracy; in Ghana, Rawlings became a civilian to run for and win the presidency; and in Nigeria, Babangida set aside the results of the General Election of 12 June 1993 to a chorus of opposition. Democracy is having a difficult second coming in Africa.

21 Further reading

Brooke-Smith, R. (1987) *The Scramble for Africa: Documents and Debates*, London: Macmillan.

Brunschwig, H. (1966) *French Colonialism, 1871–1914*, London.

Chamberlain, M.E. (1974) *The Scramble for Africa*, London: Longman.

Fagan, B. (1965) *Southern Africa: During the Iron Age*, London: Thames & Hudson.

Fieldhouse, D.K. (1981) *Colonialism 1870–1945: An Introduction*, London: Weidenfeld & Nicolson.

Hargreaves, J.D. (1974) *West Africa Partitioned*, London: Longman.

Hobson, J.A. (1902) *Imperialism: A Study*, London: Allen & Unwin.

Hibbert, C. (1982) *Africa Explored*, London: Penguin.

Johnston, H.H. (1899) *The Colonization of Africa by Alien Races*, Cambridge: Cambridge University Press.

Keltie, J.S. (1893) *The Partition of Africa*, London: Stanford

Leakey, R. and Lewin, R. (1977) *Origins*, London: Macdonald & James.

Lucas, C. (1922) *The Partition and Colonization of Africa*, Oxford: Clarendon Press.

Mackenzie, J.M. (1983) *The Partition of Africa*, London: Methuen.

Morel, E.D. (1920) *The Black Man's Burden*, Manchester: National Labour Press.

Mudenge, S.I.G. (1988) *A Political History of Munhumutapa c. 1400–1902*, London: James Currey, and Harare: Zimbabwe Publishing House.

Penrose, E.F. (ed.) (1975) *European Imperialism and the Partition of Africa*, London: Frank Cass.

Plass, M.W. (1967) *African Miniatures: The Goldweights of Ashanti*, London: Lund Humphries.

Robinson, R. and Gallagher, J. (1961) *Africa and the Victorians: The Official Mind of Imperialism*, London: Macmillan.

Schreuder, D.M. (1980) *The Scramble for Southern Africa*, Cambridge: Cambridge University Press.

Severin, T. (1973) *The African Adventure*, London: Hamish Hamilton.

Taylor, A.J.P. (1938) *Germany's First Bid for Colonies, 1884–85*, London: Macmillan.

C Political

22 The states of modern Africa

The states of modern Africa are essentially colonial creations transformed into independent states. Their boundaries, shapes and sizes are part of the colonial inheritance. Mainland Africa has forty-six independent states and two dependent territories, there are six independent island states and seven dependent island groups. The states come in all shapes and sizes.

According to the 'great game of scramble', a European power first had to establish a claim to its stretch of African coastline and was then able to declare as its legitimate sphere of influence territory directly inland. The length of coastline claimed depended on how near on either side of its own trading post was the next post of another European power. On the coveted west African coastline, trading posts were strung together like beads on a necklace and some claims were for very short lengths of coastline indeed.

The former British colony of the Gambia was based on the trading post of Bathurst, now Banjul, typically located on an island at the mouth of the Gambia river. The sphere of influence inland was defined mainly in terms of that river which was the main trading artery. The result is a state over 200 miles (320 km) long with a width of 30 miles (48 km). The former German colony, Togo, has a 44 miles (70 km) coastline but extends inland about 340 miles (545 km); its neighbour Benin, the former French colony of Dahomey, has a coastline of 62 miles (100 km) and an inland extent of 410 miles (655 km). No detailed research has yet examined the relationship between territorial shape and economic development, but such extreme shapes create strong regional contrasts which suggest a prima-facie case that they do not help development.

The total area of Africa is over 11.5 million square miles (30 million sq. km). Twenty-one states, each over 500,000 square miles (1.28 million sq. km) in extent, account for 82.9 per cent of the area; the remainder of Africa and its surrounding islands is divided between thirty-one independent states and nine other territories. One of the most pertinent facts of African political geography is that seven independent states are smaller in area than Wales, and *together* make up only one-eighth of one per cent of the total area of the continent. While thirty-one African states have an area greater than that of the United Kingdom, only two, Nigeria and Egypt, have a greater population. Despite rapid population growth in recent years ten independent African states have populations under one million. The Seychelles population would still only barely fill a capacity-reduced Cardiff Arms Park! If a third criterion of size is used, wealth, as measured by gross national product (GNP), then the

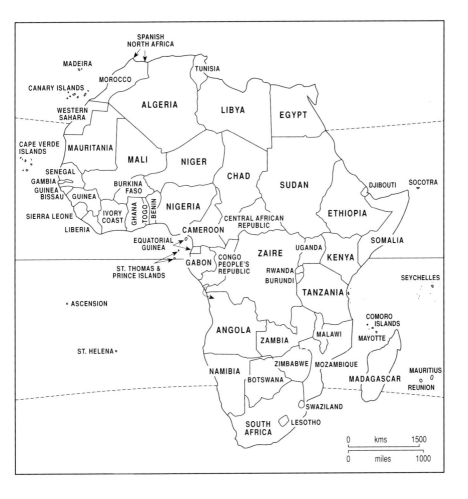

contrast between rich and poor within Africa, and between African countries and the rest of the world, is even greater. In 1990 fifteen African countries had a total GNP of less than US $1 billion and five less than US $100 million. The total GNP of all fifty-two independent states in Africa is less than one-eighth of that of the United States. For a variety of reasons, mostly originating in the colonial past, many independent African states are very small according to a number of criteria. Their extreme lack of size brings into question their economic and political viability and renders them peculiarly vulnerable to the forces of neo-colonialism.

23 Boundary problems

Africa has almost 50,000 miles (80,000 km) of international land boundaries which divide the continent into forty-six independent states, Western Sahara and the two tiny enclaves of Spanish North Africa. There are 104 different boundaries in Africa. Many, disputed in the post-independence period, have given rise to bloodshed and even war. They are a constant impediment to good international relations on the continent.

The international boundaries are part of Africa's colonial inheritance. Modern African states are, with very few exceptions, territorially identical to the European colonies they replaced, 'for all their grotesque shapes and varied sizes'. What was acceptable for colonies is often less so for independent states but, despite the obvious drawbacks of an anomolous and anachronistic political framework; opportunities to change it have largely been ignored, indeed deliberately thwarted.

The international boundaries of modern Africa emerged mainly in the thirty years following the Berlin Conference of 1884–5. That Conference, between the European powers, laid down the rules for the European partition of Africa. The boundaries were subsequently drawn by the European powers with scant regard even for the physical geography of Africa, let alone the Africans. In a series of bilateral treaties between the European powers, boundaries were drawn to define the different European spheres of influence. From *definition* the process moved through *delimitation*, which marked the boundaries on maps, to *demarcation*, which (not always completed) saw the erection of boundary pillars on the ground by joint boundary commissions. By 1914 the political map of Africa was virtually complete. The lines which remained to be drawn were: in the Sahara, internal sub-divisions of French colonial Africa, and adjustments subsequent to the First World War.

The Saharan boundaries were resolved largely by straight lines. Divisions internal to the French empire came and went; some were never completed, leaving thorny post-independence problems, notably between Algeria and Morocco. Several boundaries changed following the First World War: Jubaland was given by Britain to Italy and Kenya was extended westwards to compensate, at the expense of Uganda. Such was Italy's reward for joining the War on the 'right' side. Portugal stepped in swiftly to take the Kionga triangle from Tanganyika, having lost the 215 square mile (345 sq. km) territory to Germany only in 1909. France took the bulk of Kamerun and returned corridors of territory, which had been ceded to Germany only in 1911 to give access to the Congo (Zaire) and Ubangui rivers, to their own colony of French

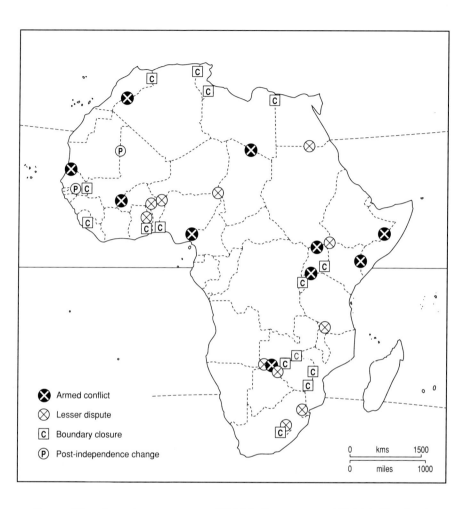

Armed conflict

Lesser dispute

C   Boundary closure

P   Post-independence change

0   kms   1500
0   miles   1000

Congo. Elsewhere the emergence of British Cameroon and British Togoland changed the western boundaries of the two former German colonies in West Africa.

Later, minor adjustments were made to suit the colonial powers: for example, an exchange took place in 1927 between Belgium and Portugal in which the 'Botte de Diolo' (2200 square miles, 3500 sq. km) passed to Angola in return for 2 square miles (3 sq. km) of the Duizi valley near Matadi, thus allowing a convenient realignment and upgrading of the Leopoldville (Kinshasa) railway.

Because the boundaries of Africa were drawn by Europeans with little knowledge of the continent and little attention to detail, there are many uncertainties and ambiguities in the 50,000 miles of lines. A watershed seems a precise enough concept until an attempt is made to delimit one, such as that between the Congo (Zaire) and the Zambesi, on a wide, almost level, plateau surface. A river is a river but is the boundary line a thalweg, median line or bank? The status of islands is unambiguous with thalweg or bank, but what if the main channel of the river changes? Does a river (or lake) bank boundary preclude a country from riparian, fishing or navigational rights?

The use of physical features (rivers and river basins, 46 per cent) and geometric lines (48 per cent) dehumanized the boundaries of Africa. Only rarely did they coincide with culture or ethno-linguistic areas. Every boundary in Africa cuts through at least one culture area. The Nigeria–Cameroon boundary divides fourteen, while the boundaries of Burkina Faso cross twenty-one culture areas. At the micro level such boundaries sometimes divide towns from their hinterlands, villages from their traditional fields; they affect everyday life.

That Africa's colonial boundaries have survived over thirty years of independence is due to the OAU, meeting at Cairo in July 1964:

> Considering that border problems constitute a grave and permanent factor of dissension ... all Member States pledge themselves to respect the borders existing on their achievement of national independence.

In the whole of Africa only one boundary adjustment was made by bilateral agreement between independent states before the OAU's Cairo accord, and only one has been made since.

In 1963 the meridian boundary between Mali and Mauritania was adjusted to another straight line to take account of the nomadic habits of tribespeople in the Sahara. In 1975 the ambiguous Anglo-French boundary at the far eastern end of the Gambia was re-defined, delimited and demarcated by joint agreement of the Gambia and Senegal. In both cases the changes were no more than adjustments of boundaries not in active dispute, but they were commonsensical and took account of local human conditions.

The OAU accord of 1964 has not prevented border disputes or frontier wars but has inhibited sensible, bilaterally agreed changes. Africa overflows with 'weak' boundaries which are time-bombs waiting for a change in political circumstance to ignite the fuse. In a change of political context, a point of weakness sometimes becomes significant and a focal point for military conflict. The mere potential for such disputes has weakened African states by apparently legitimizing military demands for scarce resources.

A new approach to encourage measured change is necessary before the new climate of political secession and boundary dispute that is currently prevalent in eastern Europe spreads to Africa. Boundary lines and nation states are alien concepts imposed on Africa. The lines were drawn with a general disregard for local human factors and an incomplete knowledge of the chosen physical features, so that problems and ambiguities, often the cause of strife, were inevitable. Such problems are best solved in Africa by Africans.

24 Land-locked states

Africa has fourteen land-locked states, as many as in the whole of the rest of the world until the break-up of the Soviet Union. They are:

Mali	Uganda	Malawi	Swaziland	Lesotho
Burkina Faso	Rwanda	Zambia		
Niger	Burundi	Zimbabwe		
Chad		Botswana		
CAR				

Most land-locked states are territories which were least integrated into the colonial systems. In the British case most were indirectly ruled protectorates such as the small kingdoms of Swaziland and Lesotho. The French Sahelian territories for much of their colonial life were ruled by the military. Rwanda and Burundi were products of First World War territorial booty. The lack of colonial interest reflected remoteness and a general absence of resources, except perhaps labour. Only Zambia, Zimbabwe and, after independence, Botswana are exceptions to this. Most land-locked states in Africa are economically and politically weak; remote backwaters dependent on stronger, richer, more accessible neighbours.

Eleven land-locked states had a GNP per caput in 1990 of under US $500 per annum. Only Botswana had a GNP per caput over US $1000. One land-locked state alone, Zimbabwe, had a total GNP over US $5000 million in 1990 and only Uganda had a population in excess of 10 million. Land-locked states also came low in the political pecking order (see Chapter 25) with, on average, a rating half that of the continental seaboard states. By most measures of size, wealth and political influence, the land-locked states as a group were among the weaker states of the continent. On the other hand, their comparative disadvantage was less in 1990 than it had been in 1981, mainly because of the relative decline of other states.

In colonial times most land-locked states were starved of investment. In particular there were few infrastructural improvements. Five land-locked states, for example, do not have a single mile of railway, while Lesotho has just that, one solitary mile. The common problem of land-locked states is that of access. Distances to the sea are often great, even by the shortest route. Chad, the Central African Republic (CAR) and Rwanda are all more than 1000 miles (1600 km) from the sea by any surface route. Burkina Faso, Mali, Niger, Uganda, Burundi, Zambia and Botswana are all more than 500 miles (800 km) inland.

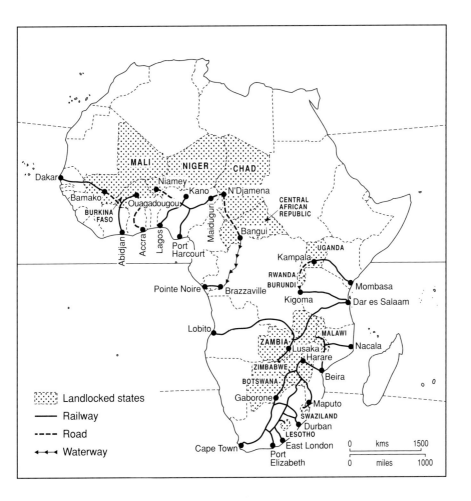

Colonial access routes were not always the shortest. The all-French route from the coast to Chad is twice as long as the most direct route and also involves costly trans-shipments between different modes of transport. It runs from Pointe Noire by rail to Brazzaville, by river to Bangui, and thence by road to N'Djamena, a distance of 18,000 miles (2900 km) through three countries. From Port Harcourt by rail or road to Maiduguri and thence to N'Djamena by road is about 1085 miles (1750 km) through two countries. The route from Cape Town to the Zambian copperbelt via the spinal railway is about 2150 miles (3440 km) compared with the route from Dar es Salaam of about 1130 miles (1810 km). The Tanzam route, now followed by oil pipeline, tarred road and railway which were all completed between 1968 and 1975 after

71

Zambia's independence, cost over £200 million to develop. Other costly post-independence routes from land-locked states include the Malawi–Nacala rail link, and tarred roads from Rwanda and Uganda to Mombasa, from Burkina Faso to Accra, and from Mali to the Abidjan railway. All imposed considerable financial burdens on poor states.

The land-locked state is vulnerable in a way in which no seaboard state is. It can be threatened by closure of a frontier, as between Zambia and Rhodesia intermittently from 1966 to 1979, Lesotho and South Africa in 1986, or Mali and Senegal in 1960. Rhodesia was itself similarly pressured in 1976 and also

African land-locked states: access routes

Land-locked state	Capital city	Port	Mode of transport	Approx. distance Miles	Km
Botswana	Gaborone	Cape Town	rail	975	1560
		Port Elizabeth	rail	800	1280
		Beira	rail	1065	1705
Burkina Faso	Ouagadougou	Abidjan	rail	725	1160
		Accra	road	725	1160
Burundi	Bujumbura	Dar es Salaam	lake/rail	900	1440
		Matadi	road/river/rail	1825	2920
CAR	Bangui	Pointe Noire	river/rail	1100	1760
Chad	N'Djamena	Pointe Noire	road/river/rail	1810	2900
		Douala	road	1175	1880
		Port Harcourt	road	1085	1735
		Lagos	road	1240	1985
Lesotho	Maseru	Durban	rail	390	625
		East London	rail	400	640
Malawi	Lilongwe	Beira	rail	460	735
		Nacala	rail	570	910
Mali	Bamako	Dakar	rail	800	1280
		Abidjan	road/rail	745	1190
Niger	Niamey	Cotonou	road/rail	710	1135
		Accra	road	880	1410
Rwanda	Kigali	Mombasa	road/rail	1035	1655
Swaziland	Mbabane	Maputo	rail	225	360
Uganda	Kampala	Mombasa	rail	725	1160
Zambia	Lusaka	Dar es Salaam	rail	1150	1840
		Lobito	rail	1650	2640
		Beira	rail	1245	1990
		Beira	road	650	1040
		Port Elizabeth	rail	1765	2825
Zimbabwe	Harare	Beira	rail	375	600
		Maputo	rail	770	1230

reacted by building a new rail link to the coast via a friendly state, in this case South Africa. The vulnerability of Zambia and Zimbabwe in the 1980s was demonstrated as rail links to the sea at Lobito and Beira respectively were disrupted as part of South Africa's destabilization of her neighbours.

A land-locked state is perhaps most vulnerable when it depends on large and regular bulk transport, as Zambia does for copper exports. Issues such as congestion of foreign ports, operational efficiency and rolling-stock deployment became vitally important. A steady flow of traffic to and from a land-locked state requires careful co-ordination and close co-operation between its own transport authority and that of the access state, in minutiae of operation and maintenance as well as in matters of investment and route development. This is rarely achieved, if only because access states have different priorities for their scarce resources.

The development of air transport has meant that no land-locked state is completely isolated. In the emergency of Rhodesian oil sanctions, supplies to Zambia were ferried in by the Canadian Air Force. Air transport is extremely useful in an emergency but is only a short-term amelioration of the problems of land-locked countries.

From the mid-nineteenth century, when Britain isolated the Boer republics, leaders of African land-locked states have developed geo-political claustrophobia, the main symptom of which is obsession with routes of access. Their concern is always with alternative routes, because the folly of relying on a single route has been underscored time and again. The special continental viewpoint of the land-locked state adds an important psycho-geographical dimension to the geo-politics of African development (refer to Chapter 58, Transport).

25 A political pecking order

Africa is a continent of fifty-two independent states, some large and many small as gauged by area, population and wealth. Each measure of size has its shortcomings, but they are sufficiently well known to cause little problem. What is not available is a measure of something less tangible, that of a nation's political status among its peers.

An attempt is made here to do just that by quantifying for each African country the number of permanent diplomatic missions that it has in other African states plus the number of other African states with a permanent diplomatic mission in residence. Because states with many immediate neighbours might be accorded a diplomatic mission more as a matter of convenience than of status, account is taken of contiguity by excluding missions to or from adjoining states. The totals are then expressed as a percentage of non-contiguous states, in order to arrive at an index (0–100) of political pecking order. The higher a state is on the index, the higher the status accorded to it by its own continent.

Egypt and Nigeria had the highest status among African states in 1991, followed by Algeria, Zaire, Libya and Ethiopia. Egypt was host to thirty-one diplomatic missions, all but one (Sudan) from non-contiguous states, and had forty missions in other African states, again all but one (Sudan) in non-contiguous states. Nigeria hosted thirty missions, three from contiguous countries, and itself had thirty-five missions in other African countries. Although at a rather lower level, Algeria and Libya also had more missions abroad than they hosted, whilst Zaire had exactly equal numbers. Ethiopia had less than half of the number of missions abroad than it hosted. The status of Egypt and Libya seems to have risen over the past decade, partly as a result of their own diplomatic initiative, whilst that of Zaire has fallen. Under the Emperor Haile Selassie Ethiopia acquired enormous diplomatic prestige within Africa and that has by no means been fully eroded. The continued presence of the OAU, the Economic Commission for Africa (ECA) and other international bodies which are influential on the continent account for the high number of missions resident in Addis Ababa. However, on the other side of the coin, the fact that fewer than half of those missions are reciprocated by Ethiopia is more in keeping with Ethiopia's poverty and political instability. The very presence of Ethiopia in the top six diplomatically expresses the purpose of this exercise because Ethiopia is one of the poorest states in the continent.

At the next level of states Senegal appears to have fallen behind Ivory

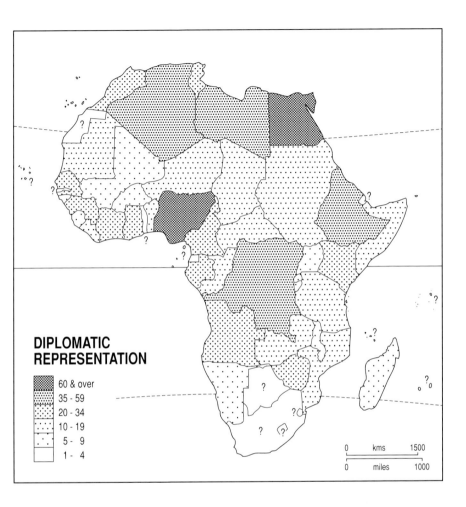

DIPLOMATIC REPRESENTATION

- 60 & over
- 35 - 59
- 20 - 34
- 10 - 19
- 5 - 9
- 1 - 4

0   kms   1500
0   miles   1000

Coast, whilst comparative newcomers Angola (1975) and Zimbabwe (1980) have advanced in diplomatic terms. In contrast, the position of Zambia has declined in these terms, perhaps because the 'front-line' moved further southwards in the 1980s to a rather richer Zimbabwe.

At the lowest levels are the poorest, land-locked, most peripheral and island states. There is a stronger correlation between political status and aggregate size of economy than status and wealth per head, as in the case of Botswana which achieves Africa's top ten in terms of GNP per caput but remains in the lowest category of political pecking order. In 1992 South Africa retained its status of pariah state within Africa, something that is likely to change over the next decade.

26 Africa must unite!

'Africa must unite!' was the unheeded clarion call of Kwame Nkrumah, the first President of Ghana, which was the first black African state to win its independence in 1957. His further prophetic words pointed to: 'the necessity to guard against neo-colonialism and balkanization, both of which would impede unity'. Thirty years after those words were first penned the first generation of post-colonial Africa has witnessed little unity but has experienced the severe deleterious effects of neo-colonialism and balkanization.

African independence was on the basis of small units. The two great French colonial federations (AOF and AEF) were divided before independence. The attempt to keep Senegal and Mali together lasted only two months before the fragile unity broke in acrimonious dispute and a closed border. In British East Africa an attempt to keep the three colonies in step by sharing the same independence date, the easier to achieve unity, was thwarted. In southern Africa the British had seen the need for larger territorial units since the 1870s, but theirs was an imperialistic concept and their two creations became vehicles to prolong white minority rule. The Union of South Africa (1910) united four colonies into a single, rich and strong state, the only one to be so created in Africa and under white minority rule into the 1990s. The Federation of Rhodesia and Nyasaland lasted less than ten years as a white settler-dominated state and did not survive independence.

Elsewhere unions of colonies were rare and did not strengthen the resultant state. Eritrea gave Ethiopia a coastline and a thirty-year secessionist war. Morocco, reunited from 1956, forcibly took Western Sahara in 1975 and holds its useful parts behind sand walls as conflict continues. The union of former British and Italian Somaliland in 1960 seems to be undone in 1992. Since the union of Tanganyika and Zanzibar in 1964 both parties have gone mostly their own way.

Once political independence was separately achieved, unity between African states became virtually impossible. Nkrumah saw the need for unity very clearly and led Ghana in 1960 into the 'Union of African States' with Guinea and Mali which, he hoped, 'would prove to be the successful pilot scheme to lead eventually to full continental unity'. Opposition came from states which were 'jealous of their sovereignty and tended to exaggerate their separatism'. In 1961 African states, though all professing the aim of 'some kind of unity', divided as to the means of achieving it. The Casablanca group, including Ghana, put political unity first, the Monrovia group gave priority to economic associations. By the time that the OAU was set up in 1963, the

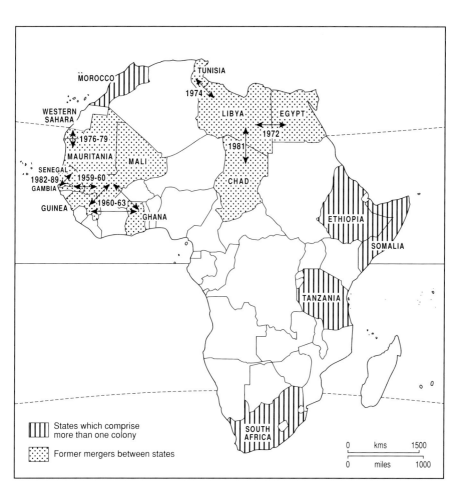

States which comprise
more than one colony

Former mergers between states

Monrovia group had won and even the Union of African States disintegrated.

The various impulsive proclamations of unity between Libya and her neighbours have led to nothing except closed borders. The apparently well-set-up East African Community, which fell far short of political union, came apart over regional economic policy, conflicting ideology and Idi Amin. The Federation of Senegambia was short-lived (1982–9) despite the Gambia being one of Africa's mini-states. Unity, which can be strength, must be based on agreed aims which take time to work out, even longer to implement. There seems to be no real prospect of any African states achieving such a unity in the foreseeable future.

27 Libya and its neighbours

Libya, under Colonel Gadafy, sometimes with no prior discussion with the prospective partner, has announced union with Egypt (1972), Tunisia (1974), Chad (1981) and non-contiguous Morocco (1984). At different times Gadafy has tried to destabilize the governments of all four states. Not an easy neighbour, and an arch-meddler in the affairs of other states, his interventions are often ill-considered, as when he sent troops to help Idi Amin in Uganda in 1979, or inconsistent in supporting non-Muslim Ethiopia against Muslim Somalia and Eritrea.

The unreliability of Libya in international politics extends beyond dealings with its neighbours. Gadafy, who came to power in a *coup d'état* in 1969, takes an independent, radical but erratic line which in 1987 goaded the American Reagan administration into bombing Tripoli. The raid caused many civilian casualties but, despite being personally targeted, Gadafy escaped. American aggression seems to have had a sobering effect on Gadafy who took a 'moderate' line over the Gulf War in 1991. Overtures to Britain to restore the diplomatic relations that were cut after the London Bureau (Embassy) incident of 1984 foundered over alleged involvement in the Lockerbie and Niger airliner bombings. Western sanctions against Libya are threatened in 1992.

Libya is a state with a small population but large oil resources, giving rise to the highest per caput income in Africa. It appears to be ruled largely at the whim of one man, who is erratic, so that his good points are constantly being undone as he seems to challenge international order by promoting violence and supporting terrorism.

Gadafy's expansionism in Chad in 1981–2 and apparent French willingness to concede to him all but *L'Tchad utile* alarmed Nigeria, which committed an armed force under the banner of the OAU. Libya withdrew, except from the uranium-rich Aouzou strip, only to invade Chad again in 1983. The French, as neo-colonial protectors of the Chad government, were persuaded by conservative Francophone African states and the United States to commit a large force to confront the Libyans, who reluctantly withdrew leaving some diplomatic egg on French faces.

Western concern about Gadafy's increasing his political influence in Africa south of the Sahara heightened when Captain Sankara came to power in a *coup d'état* in Burkina Faso in 1983. Fears of a radical Tripoli/Ouagadougou/Accra axis, which were probably greatly exaggerated anyway, faded when Sankara was assassinated in 1988, leaving Burkina Faso still under a military dictator-

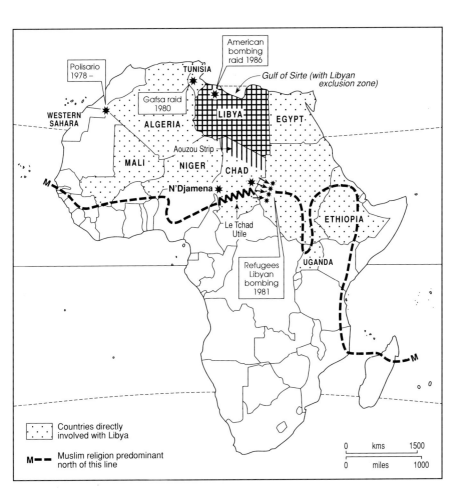

American bombing raid 1986

Polisario 1978 –

TUNISIA

Gulf of Sirte (with Libyan exclusion zone)

Gafsa raid 1980

WESTERN SAHARA

LIBYA

EGYPT

ALGERIA

Aouzou Strip

MALI

NIGER

CHAD

N'Djamena

ETHIOPIA

Le Tchad Utile

UGANDA

Refugees Libyan bombing 1981

M

M

Countries directly involved with Libya

M ━ ━ Muslim religion predominant north of this line

| 0 | kms | 1500 |
| 0 | miles | 1000 |

ship but one better disposed towards France and the United States, and not overly pro-Gadafy.

Despite having a sound, small but well endowed power-base, Gadafy has achieved little except to draw the wrath of the United States. He appears to have few international friends. Gadafy's unpredictability and his predilection to espouse terrorism have made him no more than a minor irritation, a gadfly to successive American administrations. In line with this is his bold but ill-considered assertion that the Gulf of Sirte is exclusively Libyan, backed up with almost suicidal fighter plane sorties against the might of the American Mediterranean fleet. To what effect? Africa's needs are for more considered co-operation between states and less gratuitous antagonism.

28 *Coups d'état* and military rule

Africa's penchant for the *coup d'état* continues unabated. *Coups* account for by far the greatest number of government changes in post-colonial Africa where there have been seventy-nine successful *coups d'état*. In contrast there have been forty-eight constitutional changes of government, most resulting from the death or retirement of the incumbent and only four from electoral defeat. Of the thirty-one African states which have suffered successful *coups d'état* twenty-two have had more than one *coup*.

Along with the *coups* there has been a marked increase in military leaders in power. In the five years 1965–9, twenty-one *coups* occurred in Africa and the number of military leaders in power rose from one to thirteen in that period. By 1978 there were twenty-two military rulers in Africa and in every year since 1985 a majority of African states have had military leaders, twenty-seven of the fifty-two states in 1992.

Every decade since independence has been marred by the violent death of a head of state. Thirteen African leaders have been assassinated in office, from Abubakar Tafawa Balewa in 1966 to Murtala Mohammed in 1976, Sadat in 1981, Doe in 1990, and Boudiaf in 1992. Eight have been murdered or executed following *coups d'état*.

In contrast, largely because of the ravages of time, the 'reaper' and a very little help from the ballot box, only seven states have the same leader with whom they attained independence. Two leaders from the 1960s remain in office; the nonogenarians Felix Houphouët-Boigny of Ivory Coast and Dr Hastings Kamuzu Banda of Malawi have led their countries for thirty-two and twenty-eight years respectively.

A *coup d'état*, that in Egypt in 1952 which ousted King Farouk and eventually brought Abdul Nasser to power, marked the beginning of the end of colonialism in Africa. Nationalization of the Suez Canal, the Suez crisis and the humiliation of the two great African colonial powers, Britain and France, in 1956 had far-reaching effects. But events in Egypt did not set the pattern for the spate of *coups* that swept black Africa after independence. Of the twenty-five states which achieved independence during 1957–62 nineteen have since experienced between them fifty-five successful *coups d'état*. Eleven had their first successful *coup* within six years of independence and fifteen have experienced more than one successful *coup*. Each of the three full decades since independence has seen twenty or more *coups* in Africa, and in the 1990s there have already been six.

The occurrence of so many *coups* soon after independence suggests that the

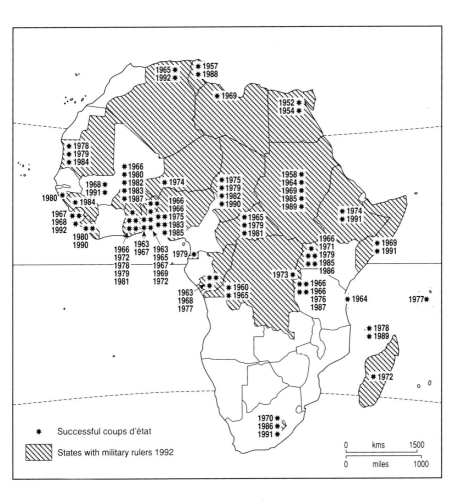

root causes existed at the time of independence. A state such as the Congo (Zaire) was ill-prepared for independence and a violent chaos ensued, its flames fanned by colonial mining interests. That lack of preparation for independence may be a factor seems to be confirmed by eleven of the fourteen former French colonies to achieve independence in 1960 having experienced successful *coups d'état*. But there seems to have been a delay factor and the tragedy of the Congo (Zaire) was the exception rather than the rule. Ten of the new states of 1957–62 experienced their first successful *coups d'état* between three and six years after independence. In some, political instability derived from a growing disillusionment with the economic reality of independence. Barely viable colonies made chronically poor states. Riches did not flow and the maldistribution of wealth remained much as before with black élites simply substituted for white colonial élites. New governments were often profligate with meagre resources and the cost of the very symbolism of nationhood was frequently excessive. Political power was abused and corruption in high places became a fact of life.

Once the unifying ideal of independence was achieved, compromises and coalitions which were made in the name of that ideal began to fall apart. Most African states had no ideology beyond independence. Carefully arranged checks and balances between different language or religious groups, between conservatives and progressives, or simply between regions, were difficult to maintain once the colonial power at the fulcrum had been removed. With the rewards for political success being considerable and at the absolute disposal of the government of the day, the temptation to seek or retain power by any means, including the unconstitutional, was overwhelming. On only four occasions since 1950 has power in an African state passed from one civilian government to a different civilian government as the result of an election. Politicians who were determined on unconstitutional means of gaining or retaining power naturally enough turned to the military for support. Sometimes this ploy rebounded when a civilian *coup* with military support was succeeded by a straightforward military *coup d'état*.

In Uganda the conservative/progressive and regional and ethnic balances that were created in the independence constitution were smashed by Obote using a then little-known army commander, Idi Amin. Also in 1966 Nwambutse IV of Burundi was deposed by his son Ntare V, who, before the year was out, was himself deposed by his military associate Micambero who had assisted in the original *coup*. In Uganda Amin ousted Obote in less than five years. Except in Morocco and Swaziland, all African monarchies and would-be dynasties have been overthrown, including those in Tunisia, Uganda and Lesotho where post-independence power was originally shared.

Regional rivalries have often led to successful *coups d'état* in African countries, notably in the Congo (Zaire), Nigeria and Uganda. These takeovers of the central government have been different from secessionist attempts as also seen in Nigeria, as well as in the Sudan, Ethiopia and Somalia.

External factors can have a destabilizing effect, as in the toppling of the long-established Ould Daddah government (1960–78) in Mauritania. Involvement in Western Sahara as the accomplice of Morocco rebounded as POLISARIO attacks disrupted the flow of Mauritania's life-blood, iron ore from the Zouerate mines. The costly war led directly to a military *coup d'état* which was immediately followed by disengagement, a policy more in line with Mauritania's traditional support of the Saharwis and distrust of Morocco. In Lesotho in January 1986, intervention from South Africa, including a tight blockade of the enclave country, led to the overthrow of Chief Jonathan in a military *coup d'état* led by Brigadier Lekhanya. Chief Jonathan had made himself unpopular with Pretoria not least by inviting North Korean army officers to help to train the Lesotho army. Lekhanya's first action on taking power was to visit Pretoria to assure South Africa of his new government's good intentions.

Outside interference or support has determined the success or failure of many an African *coup*. For every successful *coup* in post-independence Africa there has been at least one unsuccessful attempt. An accurate count of the latter is not possible because they are sometimes manufactured by an incumbent government to rally support to itself. Former colonial powers, notably Britain, France and Belgium, have often intervened either to put down attempted *coups*, as in Kenya, Tanzania, Chad and Zaire, or to support *coups*, whether directly or by instant diplomatic recognition, as in the Central African Empire (Republic) or Uganda.

The military are a major force in African politics. They are coherent, relatively well-disciplined groups who are able to stand apart from any civilian government heading for trouble. They alone have the arms capability, the independent mobility and the logistics that are often necessary for staging a *coup*. In many cases the military have seen themselves, with some justification, in the role of seizing power to 'clean up' a civilian mess. They have a sense of mission which, harnessed to military ruthlessness, has sometimes led to speedy and drastic measures to right perceived wrongs. But there is rarely a unifying ideology or even a long-term strategy. In office the originally coherent military have often split into factions, resulting in new *coups*.

The military have not been eager to relinquish power. In only two states is there now a civilian leader where once there was a military one, in six others civilian rule followed military rule only for the military to take over again. The year 1979 saw a false dawn for civilian government in Africa. Two of Africa's most grotesque dictators, Amin and Bokassa, were overthrown in Uganda and the Central African Empire (Republic) respectively, and civilian rule returned to Ghana and Nigeria. Within six years all four countries had returned to military rule.

Much emphasis is now placed on the democratization of African politics through outside pressure. But a majority of African governments are not even civilian let alone elected as part of a multi-party democracy. Military rule in Africa and the neglect of the ballot box in favour of the bullet are habits which are not easily broken down. What is more, there are plausible reasons for military take-overs in the African context as there also are for one-party political systems. Corruption among civilian politicians is all too common, and confrontational politics might be a luxury where every year is crisis year and where all available talent is needed to fight the problems of development.

29 Zaire: the threat of secessionism

The Belgian Congo (Zaire) was the African colony probably least prepared for independence and Belgium the colonial power probably least prepared for its new status. The immediate result was a chaos of terror and brutality which underlined the fragility of African independence in 1960, exposed the forces of neo-colonialism and emphasized the dangers of national disintegration through regional secession. Also, for the second time in recent history, the Congo caused Africa to become the board on which international rivalries of non-African powers could be played out with no harm to themselves.

The crisis on the Congo was complex in its causes and convuluted in its course. Political development was rudimentary, recent and regionally based. Lumumba from Stanleyville (Kisangani), the most radical and most popular leader, became prime minister and Kasavubu, the Bakongo leader, president. Independence came on 30 June 1960, in an atmosphere of distrust between the new government and Belgium.

On 5 July 1960 sections of the Congolese army (*Force Publique*) mutinied against its Belgian officers (there were no Congolese). As violence spread, Europeans fled the country. Lumumba dismissed the Belgian commander and on 8 July appointed Mobutu as army chief of staff, while Belgian paratroopers flew in to protect Belgian citizens and property. On 11 July the copper-rich region of Katanga (Shaba) was declared independent by its leader Tsombe. Lumumba appealed to the UN, Ghana and the Soviet Union for help, and a UN-led operation was mounted to enable the Congo to survive as an integral state.

The Congolese army failed to make any impression on Katanga, where Tsombe was able to deploy a white mercenary-officered army as a result of financial backing mainly from the large Belgian mining corporation *Union Minière du Haut Katanga*. In September 1960 Lumumba and Kasavubu fell out, tried to dismiss each other but were both dismissed by Mobutu in a military *coup d'état*. Lumumba escaped, was recaptured and then flown to Katanga where he was murdered in January 1961.

UN resolve to reunite the Congo strengthened but it took two years and the life of Secretary-General Hammarskjöld, who was killed in a mysterious air crash over Northern Rhodesia (Zambia), to achieve. Civilian rule returned under Adoula but further rebellion broke out at Stanleyville, Lumumba's former power-base. Mobutu ousted Adoula and invited Tsombe to accede to the Congo premiership. Katangan forces assisted by Belgians and Americans put down the rebellion. Tsombe's usefulness over, Mobutu staged a second

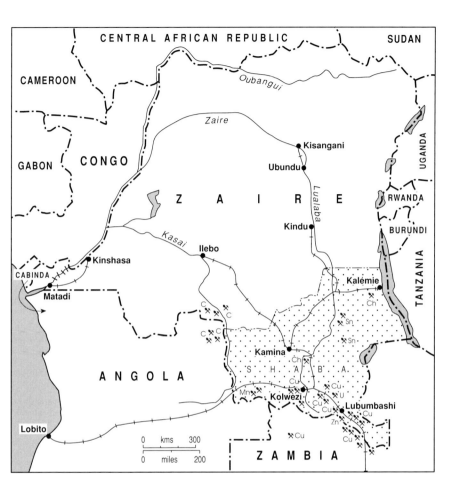

coup d'état in 1964 and has ruled ever since.

The Congo became stable but heavily dependent on American support, which was freely given to keep out the Soviet Union and to secure strategic minerals. Without policy, except 'traditionalism' in the early 1970s, the Mobutu regime has been characterized by corruption from the very top and the threat of secession has never completely gone away. Zaire served American purposes in Angola as a clandestine staging-post for arms supplies for UNITA. The Soviet demise opened American eyes to the corruption of its puppet. Amidst growing unrest Mobutu survives, afloat on Stanley Pool, procrastinating on multi-party democracy and trying to judge when it will be most expedient to fly to his Swiss bank account.

30 Nigeria: strength through unity

Nigeria survived a bitter secessionist war as a single state to emerge as a major political force in Africa with the highest population and, at one time, the largest economy. Undermining this position is diminished prosperity through oil slump and economic mismanagement, so that Nigeria's economy now ranks fourth in size behind South Africa, Algeria and Egypt; a low average GNP per caput at US $310 (though this figure may be too low, see below); and an apparent need for military rule with only ten years of civilian government in thirty-two years of independence.

Controversy surrounds the detailed enumeration of Nigeria's population. Before the first post-independence census in 1963 it was announced that the population count for each internal state would form the basis for the allocation of parliamentary seats. A political premium was placed on generous head-counting as the returns duly showed. The total population of 55.7 million was hotly disputed as being far too high because of regional inflation of the figures. A second census of 1973 recorded 79.8 million, contrasted with the UN mid-year estimate of 59.7 million (a discrepancy of over one-third), and was officially set aside. In 1978 the Federal Election Commission estimated a total population of 78.5 million, compared with the UN mid-year estimate of 74.6 million. By 1989 the UN mid-year estimate was 113.8 million, an increase of 39.8 million (53.4 per cent) over the decade. In mid-1991 the UN estimated Nigeria's population at 117.5 million, but an official census conducted in November 1991 gave a figure of 88.5 million.

The 1991 census was conducted with UN observers and the questions were carefully framed to avoid any sensitive political, ethnic or religious issues. The results should be reliable but, if they are, they undercut UN mid-year estimates by one-third (32.8 per cent). Even allowing for the long history of miscounting people in Nigeria and for the particular difficulties of this case, doubts must be cast on the credibility of UN estimates, not just for Nigeria but elsewhere in Africa and beyond. The implications are serious because the estimated population figure is the basis for so many other statistics, not least the much used GNP per caput. For example, the World Bank estimate for Nigeria in 1990 was US $310, which ranked Nigeria as thirty-second among the states of Africa. An adjustment in line with the revision in population would put the figure at US $410 or rank equal twenty-third among African states. 'Losing' 29 million people does not affect Nigeria's status as Africa's most populous state.

Such a large population in Africa inevitably means great ethno-linguistic

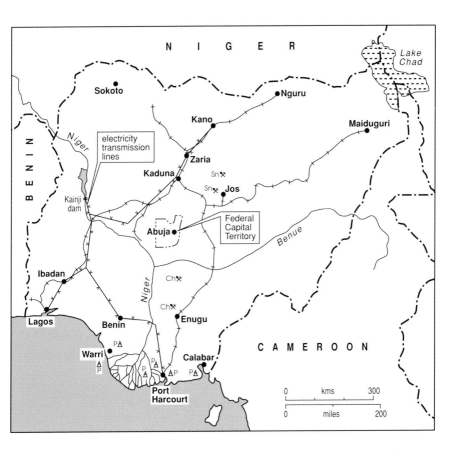

diversity. There are 395 languages in Nigeria, strictly defined as languages and not dialects. Three languages predominate: *Hausa* in the north, *Yoruba* in the south-west and *Igbo* in the south-east. The critical step from diversity to divisiveness comes not so much from the enormous number of languages but from the existence of three regionally dominant language groups.

Religion is a second major cultural divide in Nigeria, as the north is predominantly Muslim but the south is not. In the north the British practised indirect rule, reinforcing the power and Islamic conservatism of the local Emirs. In the south Christian missions made a considerable impact, not least in education, leading to a westernization and modernization of attitudes which contrasted strongly with the traditionalism of the north.

The Nigerian economy, the largest in Africa in 1980, was only fourth in size by 1990. Its rise and fall is the story of oil. In the oil boom of the 1970s the Nigerian economy expanded; as the price of oil fell through much of the 1980s, so the Nigerian economy contracted. The lesson of being so dependent upon a single resource, even oil, has been learned the hard way. Prosperity of the oil industry depends upon factors which are largely beyond Nigerian control, essentially greater or lesser supply on the world market than demand. A surplus in supply, resulting in a depressed price, has prevailed with few relaxations since 1978.

The source of Nigerian oil is the Niger delta area in the south-east of the country. The first successful well was drilled in 1956–7 before independence. Throughout the 1960s production was modest, 10–20 million tonnes per annum, but rose rapidly from 1969 to peak at 111.6 million tonnes in 1974. It stayed at over 100 million tonnes per annum until the world oil glut of the late 1970s when the Organization of Petroleum Exporting Countries (OPEC), the producers' cartel of which Nigeria is a member, set much reduced production quotas. For Nigeria the boom years of the 1970s did not last long enough and its economy was caught over-stretched with many projects only partially completed.

The boom years had created within Nigerian government circles a habit of profligacy which spelt disaster when the oil revenues plummeted. Not enough had been done in the years of plenty to diversify the economy or to go far along the path of sustainable development. The situation was not helped by the prevalence of corruption at all levels, which in turn encourages political instability to add to Nigeria's woes.

Oil had also been an ingredient in Nigeria's baptism of fire: the civil war and the attempted secession of Baifra in the late 1960s. The constitutional history of Nigeria before and after independence reflects concern with the feasibility of creating a unified state by balancing the regional interests of north, west and east, the power-bases of Hausa, Yoruba and Ibo respectively. In the run-up to independence in 1960 the emphasis was not so much on speed as on finding an appropriate constitutional formula which would stand the test of time. In the light of subsequent events, before independence it was considered that the north was the greatest threat to Nigerian unity. The independence constitution was a federal one under which the regions delegated to the centre certain powers, including control of the army, police, customs and excise, currency, central banking and regulation of international trade. The principal political parties became increasingly regionally based but independence was achieved on the basis of a united federal Nigeria.

At independence the federal government was a coalition between north and east which did not last long. A political split in the west led the federal government to take over the regional administration, eventually to install in power the minority group that was aligned with the north and to create a new Mid-West region. The 1963 census and the 1964 federal election strained north–east relations and the east was strengthened by its growing oil production. In January 1966 a pro-Ibo army *coup* killed the federal prime minister, Abubakar Tafawa Balewa, and the regional leaders of north and west. The new military leader, General Ironsi, proclaimed a military state but in July, after anti-Ibo riots in the north, he was overthrown in an anti-Ibo *coup* which installed Colonel Gowon. Reconciliation failed as further riots in the north killed thousands more Ibo who were resident there. In July 1967 the Ibo east declared the independent state of Biafra and the Nigerian civil war began. After an initial Ibo advance through the Mid-West region had been repulsed, the military outcome was never in doubt. The tragedy was that the federal forces were unable to end the affair quickly because outside powers, including

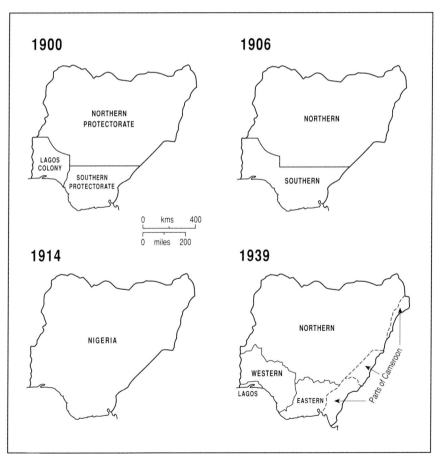

France, Portugal and South Africa, sustained Biafra until it dwindled to not much more than a single airstrip. Thousands of innocent civilians died and many more suffered severely from a war that lasted two and a half years.

The Federation of Nigeria survived intact, reconciliation began and, because of the astonishing expansion of oil production at a time of rising oil prices, a strong and rich state emerged from a sea of colonial balkanization and political and economic weakness. Despite some successes, the Gowon military government began to drift and was overthrown in a bloodless *coup* in July 1975. The new leader, General Murtala Mohammed, was vigorous and efficient, dealing with long-standing problems and detailing a timetable for a return to civilian rule. But in February 1976 he was assassinated. Government passed peacefully to General Obasanjo, with maximum continuity of purpose, and in 1979 civilian rule was restored. Shehu Shagari was elected to a new American-type, executive presidency with a fixed four-year period of office. He faced a difficult economic situation as the oil industry, the basis of the Nigerian economy, experienced the oil-glut crisis. In a crash programme of economic austerity, imports were banned and the first of several devaluations of Nigeria's currency was made. There were also problems which derived directly from the years of oil boom. The vast oil riches were enjoyed by the few who flaunted and squandered their new-found wealth. The political and commercial scene was dominated by graft and corruption. New elections in August 1983 returned Shagari to power but not without violence and deaths in Ibadan and allegations of ballot rigging elsewhere. On 31 December 1983 the army stepped in once more to overthrow the government, with the express intention of cleaning up the mess of civilian maladministration. The new military government under General Buhari clamped down on corruption but had little positive to offer and, in August 1985, was replaced in another bloodless *coup* by yet another military government, this time under General Babangida. The new government had as a priority making Nigeria face up to the economic realities of life after the oil boom. Burdened with an international debt which was actually larger than Nigeria's GNP in 1988 (almost US$30 billion), the military government applied to the International Monetary Fund (IMF) for assistance. A structural adjustment programme was imposed, the debts were rescheduled and a loan was negotiated with the World Bank. Continuing political instability was signalled by another attempted *coup* in April 1990. After several deferrals the Babangida government is committed to a return to civilian rule in 1993 but, even if this happens, the economic problems are likely to persist. Nigeria seems to be a case of a state acquiring great riches too quickly and too easily. Government was not strong enough to channel the new-found wealth into lasting improvements in infrastructure and the construction of a sound base for economic

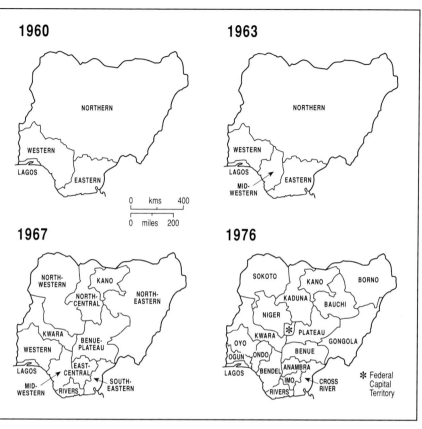

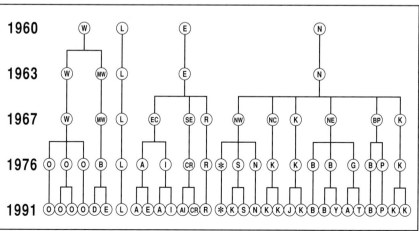

development. Wealth was squandered, corruption became a way of life, and then with cruel speed the oil industry, the source of Nigeria's riches, went into worldwide recession. If Nigeria can accomplish the hard task of recovery from the mismanagement of its all-too-brief period of wealth, the prospects for the future are bright. Some economic lessons have been learned, even if the hard way. The danger is that economic recovery will be threatened by recurring political instability.

One worrying aspect of political life in Nigeria is the inability to escape from the cleavages between north, west and east that were present before independence and still persist. The administrative divisions that were part of the colonial inheritance from the British are still part of the political map of Nigeria. The basic pattern of north, west and east was created by the British in 1939 and survived independence. The Western region had the Mid-West split off from it in 1963. There followed sub-divisions of the original three regions in 1967, 1976 and 1991, in total increasing the number of administrative areas to thirty-one. But the old regions are still identifiable and so, if there is to be any hope of ending rivalries based on the north, west and east split, a radical geographical restructuring of the administrative states of the country is needed, with new lines cutting across the old regional boundaries.

Despite the cost in lives and resources there is little doubt that the interests of all Nigerians and all Africa were best served by the preservation of a united Nigeria. Future strength depends on the ability to adjust to the new economic order. Oil helped to divide Nigeria in the 1960s, provided the basis for recovery from the civil war in the 1970s, only to be a root cause of the economic crisis of the 1980s. But oil remains a very considerable asset and, properly managed this time, could again be the platform for a more sustained Nigerian prosperity in the 1990s. But future prosperity depends on negotiating a smooth return to civilian rule. That prospect was not enhanced when Babangida set aside the General Election result of 12 June 1993, which independent observers had declared fair, and banned the presidential victor, Abiola, from a future election. Wiser counsels might yet prevail to return Nigeria quickly to civilian rule on the basis of full democracy.

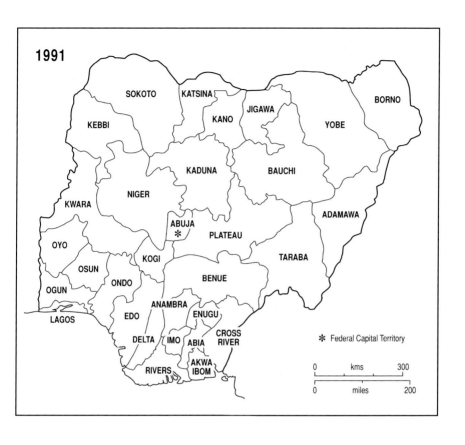

1991

SOKOTO
KATSINA
BORNO
JIGAWA
KEBBI
KANO
YOBE
KADUNA
BAUCHI
NIGER
KWARA
ADAMAWA
ABUJA
*
PLATEAU
OYO
KOGI
OSUN
TARABA
OGUN
ONDO
BENUE
ANAMBRA
EDO
ENUGU
LAGOS
DELTA
IMO
ABIA
CROSS
RIVER
RIVERS
AKWA
IBOM

* Federal Capital Territory

| 0 | kms | 300 |
| 0 | miles | 200 |

31 Uganda: ravaged Garden of Eden

Uganda, Buganda, Baganda, Muganda, Luganda: state, kingdom, people, person, language. A liturgy to confuse the outsider who is likely to be further bemused to learn that the kingdom (Buganda) is not the state (Uganda). To grasp fully the last statement is to hold the key to any understanding of the problems of modern Uganda, problems of the greatest severity which have reduced one of the most sophisticated societies in Africa to barbaric chaos and one of the richest agricultural economies to near-subsistence level.

The roots of the problems lie in the colonial past but the deadly flowering of Uganda's problems since independence was by no means inevitable, any more than was the brutality suffered by ordinary Ugandans. Nor can the whole catalogue of horror be simply laid at the door of two men, Milton Obote and Idi Amin, though they are truly culpable in their cosy exiles. Other men, be they politicians, civil servants, soldiers, police or political party thugs, by their corruption, greed, ruthlessness and cruelty are also guilty.

John Hanning Speke's 'discovery' in 1862 that the Nile flowed out of Lake Victoria opened the way to European penetration. Henry Morton Stanley's letter to the *Daily Telegraph* from Buganda in 1875 read:

> Now where is there in all the pagan world a more promising field for a mission than Uganda? ... I assure you that in one year you will have more converts to Christianity than all other missionaries united can number.

Such an evangelical trumpet call could not be resisted; certainly not by the Anglicans, who arrived hot-foot in 1877, and, despite the Anglicans' protests, by French Catholics in 1879. They vied with each other and with the Muslims for the favours of the ruler, the Kabaka. Confusion increased in 1884 when Mutesa died, to be succeeded by his sodomite teenage son, Mwanga. A reign of terror ensued with the most awful violence and bloodshed. Many Christian Baganda, both Anglican and Catholic, were martyred in the most appalling circumstances.

Captain (later Lord) Lugard arrived in December 1890 and lived precariously before establishing British 'protection' over Buganda and extending it to the other inter-lacustrine kingdoms. Even in those early days of colonial rule most of the ingredients of modern Uganda's problems were already present. They were to survive as dangerous undercurrents throughout the colonial period, to resurface after independence.

Uganda became independent in October 1962 after a troubled preparation. The difficulties arose from conflict between forces within the country which

were loosely characterized as traditional and modern and which found their strengths of expression in different regions. The traditional/modern divide was also a north/south divide and an ethnic, linguistic and religious divide. Traditional power was associated with Buganda, the largest, most prosperous and most powerful of the old Bantu kingdoms, and its hereditary ruler, the Kabaka. Modern power was vested in a new political party which commanded majority support in the country as a whole but not in Buganda; this was the Uganda People's Party (UPC), led by a northerner (Langi), Milton Obote.

In the lead-up to independence there was no question of Uganda being subdivided or, on the other hand, of a united East Africa emerging within a reasonable time-span. So, under the aegis of an enlightened British Governor, Sir Andrew Cohen, a carefully contrived constitution of checks and balances was fashioned. Within a federal structure which gave considerable autonomy to Buganda, the Kabaka, Mutesa II, was to be president and the elected majority party leader, Obote, executive prime minister. In May 1966 Obote used the army to move against the Kabaka who fled into life-long exile. The delicate checks and balances were rudely swept aside by an army which was largely northern-based and was led by northerner and Muslim, Idi Amin.

Obote became executive president under a new constitution. He consolidated his hold on the army, strengthened it, and introduced more radical measures for Ugandan development. He also devoted time to the wider issues of East African unity (the East African Community was created in 1967) and to African and Commonwealth affairs. While Obote was away at the Commonwealth Prime Ministers' Conference in Singapore in February 1971, he was deposed. A military *coup* led by Amin succeeded with limited bloodshed. The new regime was recognized by Britain with indecent haste, perhaps not unrelated to Obote's own *coup* of 1966 and to the criticism that he was currently voicing of the new British Conservative (Heath) government's attitude towards arms sales to South Africa. The fall of Obote, the man who had overthrown the Kabaka, was similarly welcomed in Buganda even though Amin was also a northerner. The tunes soon changed.

In 1972 Amin began to expel all Uganda Asians. Most Asian families had come to Uganda in the first place under British rule, from the beginning of the century, to work as traders or on the Uganda railway. It was to Britain that they now had to go and Amin's brief honeymoon with the British government and press came to an abrupt end. But Amin became an African folk hero, even president of the OAU. That popularity was also short-lived as he systematically eliminated all opposition in Uganda with frightening brutality. In his reign of terror virtually a whole élite disappeared, either into exile or wiped out. Thousands were killed, their bodies thrown into the Nile or Lake Victoria or simply dumped in a forest off the Kampala-Jinja road. The economy began to disintegrate, helped by drought in the north. With increasing religious overtones Amin acquired a Palestinian bodyguard and a Libyan military force. Uganda even produced another Christian martyr in the person of Archbishop Luwom. Amin's delusions of grandeur grew and, among other manifestations, found expression in a desire to expand Uganda's territory.

One of the curses of geo-politics is that it fascinates dictators, and Amin was no exception. Conscious of Uganda's land-locked status, he proposed a Ugandan corridor to the sea between Kenya and Tanzania and then committed himself to action on Uganda's southern boundary. It led to his downfall. The British and Germans had agreed a straight-line boundary which cut across the Kagera river, leaving a large piece of territory north of the river in Tanzania and a much smaller piece of territory south of the river in Uganda, respectively known as the Kagera salient and triangle. Each had traditionally been part of the other's territory, as the Kagera river is a natural divide. Not the first dictator to seek his country's natural frontiers, Amin invaded the Kagera salient in order to 'restore' Uganda rule. Somewhat to his surprise, the Tanzanians hit back and did not stop at the boundary parallel.

Julius Nyerere had always been implacably opposed to Amin. As a personal friend of Obote, he had given him political asylum and had encouraged other Ugandan exiles in Dar es Salaam. While Amin's brutal excesses took their toll on Uganda, Nyerere determined to intervene. The Kagera dispute triggered a well-prepared position. The Ugandan army was in disarray from continuous purges and from lack of supplies deriving from the economic crisis. At a critical moment Kenya cut Uganda's link with the sea and oil supplies. Libya sent more support but to no avail. Amin fled the country, first to Libya and then to Saudi Arabia where he still lives in quiet luxury. The Libyans were forced home and the Tanzanians held Uganda.

Provisional civilian governments, first under Youssef Lule, former vice-chancellor of Makerere University, and then under Godfrey Binaisa, did not last long. Elections in late 1980, the first in Uganda since independence, brought Milton Obote back to power. It was not a successful return. The election result was disputed and, despite the presence of Commonwealth observers, there were many unexplained irregularities. The economy was in total disarray, law and order had broken down, there was severe drought in the north and the remnants of Amin's army rampaged about the country. Above all, ethnic tensions came to the fore with factional in-fighting in the army and in the reconstituted UPC. Dissident groups took to arms and their presence in Buganda was used as an excuse for violent excesses by a Ugandan army which was increasingly ill-disciplined and beyond Obote's control. Conditions in Uganda were worse than ever and in Obote's second coming as many people were killed as under Amin's longer reign of terror. Nyerere's efforts in ridding Uganda of Amin were undone by his eagerness to reinstate Obote, who aroused all of the former divisiveness and was simply unable to cope.

In 1985 Obote was overthrown for a second time by a military *coup d'état*. However, Brigadier Okello's government could not withstand the National Resistance Army, led by Yowerri Museveni, which controlled much of the south and west of the country. In January 1986 Museveni's army entered Kampala and he was sworn in as the new president of Uganda.

More than six years later, despite valiant efforts, all of Uganda is still not fully under government control. The toll of more than two decades of civil war and brutish governments is immense. The economy is in a ruinous state and the retribution of AIDS, which affects a larger proportion of the population than in any other state, is devastating. Uganda was, as recently as the 1960s, a Garden of Eden. Can it ever be again?

32 Liberia: a civil war spills over

Liberia is a black America in Africa. Its name is synonymous with raw material exploitation, an overthrown creole dynasty, disastrous civil war which threatens neighbouring states, and rare international armed intervention. Liberia in the 1990s is a mess, where ordinary people suffer horribly or flee as refugees across the international borders. Youthful private armies brutally rampage and no solution is in sight.

As its name implies, Liberia was founded as a home in West Africa for freed American slaves. An independent state since 1847, it is the only state in Africa not to have been a European colony at some time or other. From its foundation Liberia has been very much under the wing of the United States, both politically and economically. Until 1980 it was politically stable, its internal politics dominated by a small élite group of Westernized descendants of freed slaves who lived along the coast mainly near the capital Monrovia. They kept their distance from and lordship over the diverse tribal groups of the interior.

The economics of Liberia were characterized by the exploitation of raw materials by large American-based, multinational corporations. In 1926 the major American tyre manufacturer Firestone started rubber plantations in Liberia in an enterprise so large that it dominated the Liberian economy. Iron ore deposits were later exploited, also by big American companies who built railways from the coast to inland mines in the Bomi Hills, the Bong Mountains and Mount Nimba. The railways, not connected with each other, were of different gauge. The blatant exploitation suited the ruling True Whig Party who profited greatly.

Less happy about the state of affairs were the people of the hinterland where there was little economic progress and where the basic resources were being plundered. In 1980s a bloody *coup d'état* led by Master-Sergeant Doe seized power from President Tolbert who was assassinated. Mass executions of former government members and other members of the élite group followed in a long and systematic blood bath. Samuel Doe only slowly gained recognition from other states and, despite a new constitution under which he became the unanimously voted in president, ruled somewhat precariously and erratically.

At the end of 1989 a major rebellion was started in Nimba Province in the north-east of the country, led by Charles Taylor. Thousands died and many more fled across the frontier into the Ivory Coast. Another rebel group under Prince Johnson challenged both the government and Taylor. Both groups of

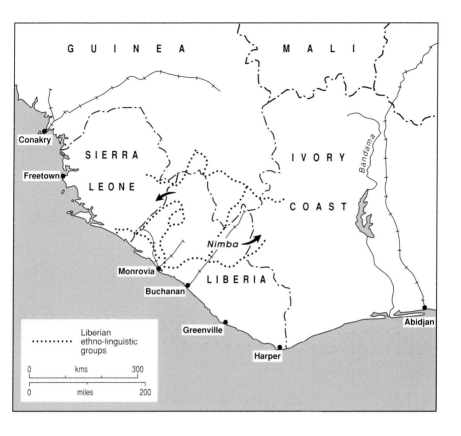

Liberian
ethno-linguistic
groups

| 0 | kms | 300 |
| 0 | miles | 200 |

rebels entered Monrovia but were not strong enough to overthrow Doe. In 1990 the regional organization ECOWAS (Economic Community of West African States) sent an international armed force to Monrovia to try to broker a peace. In the confusion of a Monrovia occupied by four armies, led by Doe, Taylor, Johnson and ECOWAS, Doe fell into rebel hands and was slowly killed. The matter did not end there. In mid-1992 Taylor held about two-thirds of the country and Johnson one-third. In Monrovia an 'interim government', under the protection of the ECOWAS force, in fact governs very little. Loyalties are often tribally based and the country is being torn apart. Worse, the conflict and refugees have spread over international boundaries, which cut culture groups, into neighbouring Sierra Leone and Ivory Coast. The havoc created holds an uncertain future.

33 Somalia: irredentism to civil war

Somalia is one of Africa's poorest countries yet it has chosen for much of its existence to sacrifice economic progress for the ideal of Somali self-determination. When that unifying ideal has not provided the driving force Somalia has degenerated into internecine warfare between the Somali clans which, in the 1990s, has reached such a pitch as to threaten disintegration of the state and has caused untold suffering among ordinary people.

At independence in 1960 Somalia was already a rarity in Africa, a union of two former colonies, British Somaliland (68,000 square miles/174,000 sq. km, 650,000 people) and Italian Somaliland (178,000 square miles/456,000 sq. km, 1,230,000 people). Even then, almost one million Somalis lived outside the newly formed Somali Republic, occupying areas of Ethiopia, Kenya and Djibouti (French Somaliland totalling about 128,000 square miles 328,000 sq. km). Roughly one-third of all Somali people and one-third of all land occupied by Somali people lay outside the Somali Republic. While other African states have struggled to create national unity from groups of diverse ethnic origin, language and religion, Somalia has been absorbed for much of the time with its aim of uniting all Somali people in a 'Greater Somalia'.

Whenever progress towards that irredentist aim has been halted, the Somalis have turned in on themselves, with the various Somali clans engaging in bitterly fought rivalry with each other for supremacy within the desperately poor country.

Either way, the people of Somalia have lost and lost very heavily. Somalia's irredentist aim and the total rejection of it by the neighbouring states has been an intractable problem exacerbated by the strategic position of the Horn of Africa which has attracted the attention of imperial powers for a century and a half. Thousands have been killed in the cross-border warfare that has punctuated the post-colonial period, thousands more have died of related cause and literally millions have become refugees. Warfare between the Somali clans has been even more bitter and every bit as destructive, as warlords with well-equipped private armies have fought each other for political supremacy. Recently it has seemed that almost all of the Somali leaders have had their fingers pressed hard on a self-destruct button, since all outside efforts to help to ameliorate a desperate situation have been rudely brushed aside. All of this takes place in an area of great natural poverty, a drought-stricken tract where, under the most favourable of human conditions, it is difficult to eke out a living.

Over two-thirds of Somalis are pastoral nomads living in a harsh semi-arid environment. The migratory patterns of such a people present a dynamic force

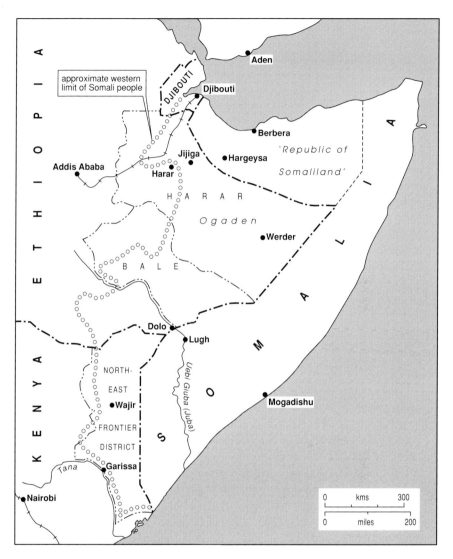

which is inevitably at odds with fixed international boundaries, especially so when those frontiers are arbitrarily drawn straight lines. For over a century Somalis are known to have migrated westward across north-eastern Kenya, displacing other groups as they moved. In the north Somalis have migrated westward into Djibouti, and the Ogaden area of Ethiopia is criss-crossed by well-defined Somali seasonal migrations.

During the 'scramble for Africa' Somaliland was divided between the spheres of influence of Britain and Italy. In 1924 Britain gave Italy Jubaland, an area of 36,740 square miles (94,050 sq. km) entirely inhabited by Somalis, as a 'reward' for Italy entering the First World war against Germany as spelt out in the secret Treaty of London, 1915. This action went some way towards solving the Somali minority problem in Kenya though that was not its main aim. Britain did make a proposal in 1946 for the creation of a 'Greater Somalia' that would have included the Ogaden region of Ethiopia. The United States and the Soviet Union, suspicious of British intentions in the strategic Horn of Africa, rejected the proposals. In 1950 the Italians were handed back their colony as a UN Trust Territory and in 1955 Ethiopia was allowed to re-occupy the Ogaden.

Between the independence of Somalia in 1960 and that of Kenya in 1963 the British were pressured to cede the remaining Somali areas of Kenya to Somalia. Although nothing came of the negotiations a new North-Eastern Region of Kenya was created in 1963, prior to the Anglo-Somali Conference in Rome where again no agreement was reached as Somalia pressed its claim to the whole of the former Northern Province of Kenya. In December 1963 Kenya achieved independence, with the 1924 boundary with Somalia intact. In July 1964 Somalia refused to sign the Cairo boundary accord of the OAU.

Kenya's independence constitution protected regional, including Somali, interests but this was swept aside in the constitutional reform of 1965. A guerilla war developed along the Kenya-Somali border. The cause of Somali self-determination received a further set-back in French Somaliland in 1967 when a referendum, for which the French carefully mobilized the non-Somali vote, declared against independence. The French changed the name of their overseas territory from *Côte Française des Somalis* to *Territoire Française des Afars et des Issas*, a title that accurately indicated French political intentions.

The superpowers, having blocked British plans for Somalia, were not long in establishing themselves in the strategic Horn of Africa. In return for a Red Sea base, the United States provided Haile Selassie with vast quantities of arms to help to keep his fragile Ethiopian empire together. The Soviet Union outbid the West to supply Somalia with weapons in return for a base at Berbera. Between them the superpowers flooded Ethiopia and Somalia with modern weaponry.

Late in 1967 Prime Minister Egal's new Somali government recognized the impoverishing effect and apparent futility of pursuing irredentism and attempted to negotiate a settlement with Kenya and Ethiopia. The guerilla war died out and Somalia was, almost for the first time, at peace with its neighbours. But removing the unifying external issue led to internal political strife which culminated in the military *coup* which brought Siad Barre to power. However, the Ethiopian *coup* of 1974 and the hype of publicity which accompanied Siad Barre's chairmanship of the OAU once more brought Somali irredentism to the fore. The Soviet Union, Somalia's superpower ally, now replaced the United States in the counsels of Ethiopia and attempted, without success, to mediate between the two sides on the self-determination issue.

Somalia moved to take advantage of the chaos in Ethiopia, and by the latter half of 1977 full-scale war was in progress. The Somalis were initially highly successful but the Ethiopians regrouped and, equipped by the Soviet Union and 'advised' by 15,000 Cubans, inflicted a crushing defeat on the Somalis at Jijiga in March 1978. The formal war was over but Somali guerilla activity, Ethiopian air-raids, and Somali cross-border incursions rumbled on.

The war coincided with serious drought to drive the Ogaden Somalis from their traditional pastures. Internationally funded refugee camps were set up along the Somali–Ethiopian frontier. Emergency food supplies were insecure, health hazards multiplied with the drought, whilst flare-ups of the war caused upsurges in the number of refugees and impeded relief work.

By 1982 Ethiopia had reasserted its full control over the Ogaden and Ethiopian forces then mounted cross-border raids on Somalia, which turned to the United States. The superpowers thus completed their musical-chairs in the Horn, each moving to support the side opposite to that which they had initially backed.

Inter-clan rivalry among the Somalis threatened and finally overcame the government of Siad Barre in 1991 and a new interim government was set up under Ali Mahdi Mohamed. But the fighting did not cease. The former British Somali-land declared itself independent, as the Republic of Somaliland, but was not recognized. In and around Mogadishu, the capital, inter-clan fighting was par-ticularly fierce through much of 1991 and 1992. International aid organizations were forced to flee an impossible situation and food aid and other supplies were unable to get through even by sea as the port of Mogadishu was continually under fire. Belatedly the UN and the United States, tried desperately to avert the worst of the crisis.

An end to the fighting, although necessary, would be only the first step in attempting to solve the problems of Somalia. Reports speak of enormous damage to almost all structures in and around Mogadishu. People are dying in their thousands because food and medical supplies cannot reach them. Basic services, including urban water supplies, have been destroyed. Beyond the crippled city drought grips the countryside and thousands are at risk of starving to death with no hope of food aid. The rift with the secessionist north will have to be healed. Beyond Somalia itself the basic problem of Somali irre-dentism remains unsolved. Superpower suspicion prevented an early solution in 1946. Superpower rivalry exploited Somali aspirations and Ethiopian disarray. Superpower arms brokers supplied the weapons for the Somalis to shoot off their own feet.

34 Angola: a cold war killing field

Potentially Angola is one of the richest states in Africa. It produces 500,000 barrels of oil per day, has reserves of almost 5 billion barrels, and has an annual production of about 1 million carats of diamonds. Other mineral resources are considerable and include iron ore, manganese, copper, nickel, gold and silver. But it is a country that has been wracked by warfare since the 1960s. The struggle for independence from the Portuguese degenerated into civil war, which was then taken over by the old superpower rivals who, turning Angola into a cold war killing field, slugged it out by proxy. The 'new world order', followed by the collapse of the Soviet Union, led to the withdrawal of the proxies, South Africa and Cuba. Angola's future is currently in the melting pot as reconciliation at the ballot box between the internal parties is attempted. There is fear of a reversion to the civil war between rival factions which are hardened and grown old in enmity.

Although Angola's independence was assured by the 1974 revolution in Portugal, in March 1975 accord between the three rival liberation groups broke down and civil war began. In the north was the *Frente Nacional de Libertacao de Angola* (FNLA), based among the Bakongo and backed by their cousins across the Zaire border. The *Movimento Popular de Libertacao de Angola* (MPLA) was a largely urban-based radical group among detribalized *assimilados* and mulattos around Luanda. In the south-east was the *Uniao Nacional para a Independencia Total de Angola* (UNITA), based among the Ovibundu people. The FNLA and UNITA were backed by the United States, the MPLA by the Soviet Union, but this backing had changed in the recent past and African states had divided loyalties.

Such uncertainties were quickly resolved when white South Africa, backed by the United States, invaded Angola in support of UNITA. The South African interest was to secure their rule in Namibia by denying SWAPO guerillas the use of Angola. Their advance was halted by the MPLA, equipped by the Soviet Union and 'advised' by thousands of Cubans. The hitherto uncommitted African states, led by Nigeria, rushed to recognize the MPLA as the Angolan government.

The South Africans retreated to just north of the Namibian border whilst UNITA held the south-eastern quadrant of Angola. From there attacks on SWAPO tightened South Africa's grip on Namibia, and attacks by UNITA on the Benguela railway and the diamond mines crippled the Angolan economy and forced Zaire and Zambia to use South African rail routes. Southern Angola became a scene of human devastation.

Again change in Angola came largely from outside. World superpower rivalry ended. South African determination to stay in Angola faded under the pressure of black unrest at home, white opposition to the war and anti-apartheid sanctions. In December 1988 an agreement on Namibia and a linked withdrawal of the Cubans from Angola was reached. By March 1990 Namibia had achieved independence and South Africa had left Angola. In 1991 a ceasefire between the MPLA and UNITA was agreed and nationwide multi-party elections were held in October 1992. President dos Santos and the MPLA were returned to power but hopes of a fairy-tale end, with a new start on the daunting task of rehabilitation of war-weary people and reconstruction of a devastated infrastructure and economy, seemed premature as UNITA refused to accept the result and hostilities became, in 1993, more widespread than ever.

35 The French connection

Much of the motivation for French colonization in Africa came from a need to boost national pride following defeat in the Franco-Prussian war in 1871. National pride and cultural arrogance was evident in France's attitude towards the administration of those colonies and towards their independence in the 1960s and now is seen in the special French brand of neo-colonialism that is experienced by the francophone states of Africa. It is perhaps these features in the relationship that make the economic advantages which accrue to France less obvious and more tolerable. At its worst the French connection with Africa is blatant neo-colonialism but, because the French are interested, active and positive in Africa, their excesses are accepted and the francophone states of Africa are mostly willing partners who eagerly crowd under the large French umbrella. In other words, *because* they interfere and assert themselves, even to their own advantage, the French are tolerated.

French influence in Africa is based on twenty former colonies and three Indian Ocean territories which remain French. Francophone influence also extends to the three former Belgian colonies of Zaire, Rwanda and Burundi and to two former British colonies, Mauritius and the Seychelles, which had two colonial languages (English and French) having passed from France to Britain in the colonial period. French 'community' with Africa is different from the British Commonwealth. The French are much more paternalistic, perhaps arising from the contrast in size and power between France and the francophone African states and the absence from the grouping of a Canada, Australia or even India and Nigeria as in the British Commonwealth.

Thirteen African states belong to the Franc Zone, with currencies linked to the French franc at a fixed rate of exchange and freely convertible into French francs. The financial reserves of these countries are held mainly in French francs and exchange is arranged through the French money market. For most Franc Zone countries in Africa France is the main trading partner, especially for imports of manufactured goods. Exports are mainly raw materials, often minerals. The trading relationship, very much in France's favour, provides an economic rationale for French interest in Africa. In countries such as the Ivory Coast much of the civil service is still staffed by French nationals.

Military involvement is perhaps the most distinctive part of the French relationship with Africa. All of the francophone states, except Guinea, have military assistance from or defence agreements with France. There are French military bases in Chad, Djibouti, Gabon, Ivory Coast and Senegal. The French have never shrunk from armed intervention in the affairs of African

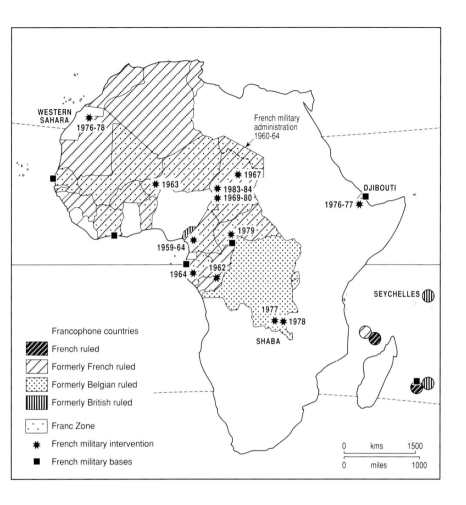

WESTERN SAHARA ✸ 1976-78

French military administration 1960-64

1967 ✸

1963 ✸

1983-84 ✸
1969-80 ✸

DJIBOUTI ■

1976-77 ✸

1959-64 ■

1979 ✸

1964 ✸ ■ 1962 ✸

SEYCHELLES ◍

1977 ✸ 1978 ✸ ✸

SHABA

Francophone countries

French ruled

Formerly French ruled

Formerly Belgian ruled

Formerly British ruled

Franc Zone

✸ French military intervention

■ French military bases

| 0 | kms | 1500 |
| 0 | miles | 1000 |

states, usually at the request of favoured African leaders to maintain the status quo, as with Gabon in May 1990 when the French sent in troops 'to protect the interests of French citizens' and Bongo retained power. However, when deemed to be in France's interest, action has been taken to oust a troublesome African leader, such as Bokassa of the Central African Republic in 1979. When his French-assisted successor (also predecessor) Dacko proved inadequate, French support was switched to a new military leader, Kolingba, who in 1981 staged a successful *coup d'état*. Chad is another state where the government has changed at the French whim, as in the overthrow of Habré, who did toe the French line, by Deby in 1990.

36 Kamerun, Cameroun, Cameroon

Germany, France and Britain at some time ruled parts of Cameroon and its boundaries changed more than those of any other territory. It occupies Africa's middle ground, richer than most but not in the top ten, and is of middling political status. Its tetchy government of over ten years moves from one minor crisis to another and niggles, or indeed nibbles away, at its neighbours.

As a colony Kamerun was originally German. Large portions of the colony were made over as concessions to two German companies, *Gesellschaft Nordwest-Kamerun* and *Gesellschaft Sud-Kamerun*, mainly for exploitation of the products of the tropical forest. The German government built two short railways from the port of Douala but did little more to advance economic development. In 1911 Kamerun was greatly enlarged by a deal between the Germans and French whereby the French gave to Kamerun two large tracts of land to the south and east, including corridors of access from Kamerun to the Ubangui and Congo (Zaire) rivers. The Germans gave France a piece of territory in the north and, most importantly, 'a free hand in Morocco'. The German corridors are not as well known as the Caprivi strip on the map of Africa because, after the First World War, most of Kamerun was given as a League of Nations mandate to France. The parts of the 1911 Convention which gave territory to Kamerun were rescinded. A strip of Kamerun, bordering on Nigeria, was given as a mandate to Britain.

The French Cameroun became independent on 1 January 1960. In British Cameroon a UN plebiscite was conducted to choose Cameroun or Nigeria. To complicate the situation the South Cameroon voted to join Cameroun, North Cameroon to join Nigeria.

The new, officially 'but only nominally' bilingual, federal Cameroon state was set up on 1 October 1961. In 1972 the federal state became the United Republic of Cameroon. Prospects were improved with the discovery of oil in the late 1970s. The first president, Ahidjo, retired in 1982 and handed over power to his colleague, Biya. Almost from that moment the politics of Cameroon have not been on an even keel. There have been attempted *coups*, regional riots and sometimes brutal suppression. Cameroon is run as a single-party state, Biya resists all outside pressure to change and sits in authoritarian unease at the centre. Relations with Nigeria are fraught, with cross-border incidents in the Cross river area routine, the initiative always coming from Cameroon. The low-level political instability hampers economic development and change at the top is probably necessary before real progress can be made. Perhaps in this way, *too*, Cameroon is also a typical African state.

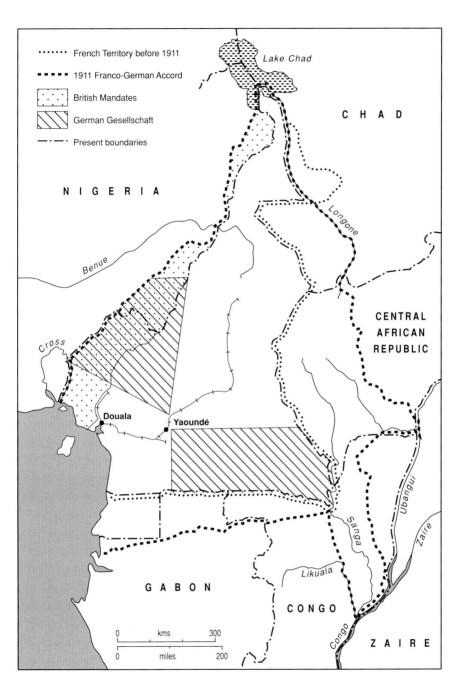

	French Territory before 1911
	1911 Franco-German Accord
	British Mandates
	German Gesellschaft
	Present boundaries

Lake Chad

C H A D

N I G E R I A

Longone

Benue

Cross

CENTRAL
AFRICAN
REPUBLIC

Douala

Yaoundé

Ubangui

Zaire

Sanga

Likuala

G A B O N

C O N G O

Congo

Z A I R E

| 0 | kms | 300 |
| 0 | miles | 200 |

37 African imperialism: Ethiopia and Eritrea

Ethiopia came to be regarded by the European powers towards the end of the last century as their equal in imperialism. Ethiopian imperialism has outlasted direct European imperialism in Africa and, despite dramatic changes in domestic political ideology, Ethiopia is still a loosely knit empire which has not yet succumbed to the forces of disintegration that have devastated much grander empires elsewhere. The main victim of Ethiopian imperialism is Eritrea but other parts of the country, including Tigray in the north and the Ogaden in the east, have recently fought militarily for greater autonomy if not downright secession.

For thirty years Eritrea fought for its independence from Ethiopia. Eritreans view the long-lasting conflict as a fight for the basic human right of self-determination, denied to them in the past by the UN. Eritrea was seen as a separate political entity which was forced into federation and then union with Ethiopia. Ethiopia has regarded the conflict as a secessionist war waged by a rebellious region which, if successful, would have left Ethiopia land-locked and in danger of further disintegration. The two positions were irreconcilable but, on the fall of the Mengistu regime in Addis Ababa in 1991, Ethiopia ended the war against Eritrea.

Eritrea contains within its colonially drawn boundaries a wide diversity of landscapes, peoples and cultures. It comprises a narrow strip of land along the Red Sea coast, over 600 miles (1000 km) long, which widens in the north to include a high plateau extension of the Ethiopian highlands and beyond that a western lowland bordering on the Sudan. *Tigrinya* speakers, who live on the plateau, make up about half of the population of Eritrea and are mainly Christian. They share their language and culture with their neighbours in the Tigray province of Ethiopia. *Tigre* speakers of the western lowland and the northern coastal strip make up about one-third of the population and are Muslim. In the southern coastal strip the *Danakil* are Muslim nomadic herdsmen, related to the *Afar* of neighbouring Djibouti.

Eritrea knew no unity before Italian colonization. From the sixteenth century the western lowland and northern coastal strip were part of the Ottoman Empire, which was succeeded in the nineteenth century by Egypt and in turn by the Mahdist state. Ethiopia held the allegiance of the plateau area but until the nineteenth century showed little interest in the coast, being for most of recent history a land-locked Christian empire dependent on its highlands for isolated survival from surrounding Islam.

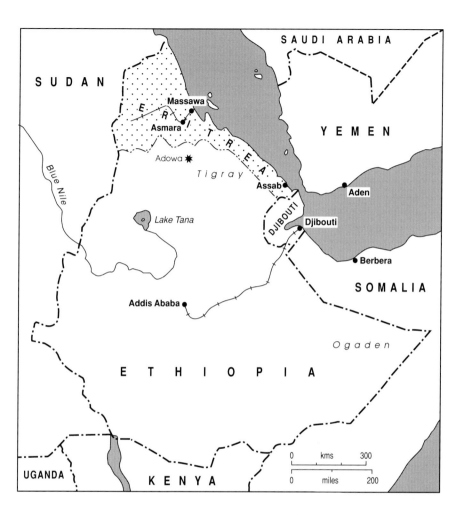

During the European scramble for Africa the ports of Assab and Massawa became Italian colonies in 1882 and 1885 respectively, and in 1890 they were incorporated into the newly formed Italian colony of Eritrea which included the whole of the coastal strip between British Sudan and French Somaliland. The boundaries of the new colony were, as usual, drawn by the Europeans. Even the name Eritrea (Erythrea) was derived from the classical name for the Red Sea. An Italian attempt then to declare a 'protectorate' over Ethiopia was defeated at the battle of Adowa in 1896. The Italians retreated to Eritrea to brood over their defeat for forty years. During that period Eritrea was, for the first time, welded into a single political entity with unified political and social structures which cut across traditional divisions.

Under Mussolini a modern Italian army conquered Ethiopia. Between 1936 and 1941, as part of the Italian East African Empire, Eritrea, along with Italian Somaliland, was ruled together with Ethiopia for the first time.

After the war Eritrea's future status had to be decided, like that of the other Italian colonies Somaliland and Libya but not Ethiopia, by a Four Power Commission of Britain, France, the Soviet Union and the United States. The Commissioners could not agree and so passed the issue to the UN, who set up a Commission of Burma, Guatemala, Norway, Pakistan and South Africa. Again the Commission was divided. Partition was rejected outright, Guatemala and Pakistan proposed the standard formula of UN Trusteeship leading to independence, but the majority favoured close association with Ethiopia. Burma and South Africa favoured federation with some autonomy, Norway wanted full union. The United States backed federation and, with only nine votes against (including that of the Soviet Union), UN Resolution 390A of December 1950 was passed. From September 1951 Eritrea became an autonomous territory federated with Ethiopia. The preamble to the resolution referred to Ethiopian claims on Eritrea, 'based on geographical, historical, ethnic or economic reasons, including in particular Ethiopia's legitimate need for adequate access to the sea'. It expressed a desire 'to assure the inhabitants of Eritrea the fullest respect and safeguards for their institutions, traditions, religions and languages as well as the widest possible measure of self-government'.

Within Eritrea there emerged a Unionist party, based in the highlands, and an 'Independence Bloc' of parties broadly favouring independence. Ethiopia, allowed great latitude by Britain to influence affairs in Eritrea, financed the Unionists and intimidated the Independence Bloc with a terrorist campaign against its leaders and supporters. An alliance between the United States and Ethiopia, who concluded a joint Defence Pact in 1953 proved decisive. British sources at the time were of the opinion that a majority of Eritreans would have voted for independence, but they were never given the opportunity.

Ethiopia consistently abused the terms of the UN Resolution and systematically set about turning federation into full union. *Amharic* became the official language and the 'autonomous' government was blatantly interfered with. Elections were held without UN supervision and a puppet regime was installed to vote for union with Ethiopia. The absorption of Eritrea excited little outside interest as the matter was considered internal to Ethiopia, which at this time commanded considerable prestige. The feudal emperor's autocratic style impressed in foreign affairs. He became the father figure of the first decade of African independence, an African who had triumphed over colonialism, whose pride and dignity had shamed the conniving politicians of pre-war Britain and France as well as the strutting Mussolini. Haile Selassie secured for Addis Ababa the headquarters of the UN Economic Commission for Africa (1958) and the OAU (1963) and with them endorsement for his government and all of its works.

The war in Eritrea escalated into fully fledged guerilla warfare on the one hand and massive retaliation on the other. Almost inevitably the Eritreans divided, the more radical Eritrean People's Liberation Front (EPLF) challenging the Eritrean Liberation Front (ELF) and both indulging in internecine warfare. The Eritreans in general were portrayed as left-wing Muslim dissidents who, by attacking conservative Christian Ethiopia, undermined United States strategy for the whole Middle East, which centred on the survival of Israel. However, in the Ethiopian revolution of 1974, Haile Selassie was overthrown and a neo-Marxist military government was installed in his place. Ethiopia turned to the Soviet Union. With regard to Eritrea, the new government was every bit as imperialistic as the old Emperor and the situation remained essentially the same. By the end of 1977 the Eritreans had gained control of all of the territory except for some garrison towns but, instead of nego-tiating with them, the Mengistu regime, now backed by the Soviet Union and Cuba, sought a military solution. In 1978 an Ethiopian army of over 100,000, with Cuban and Soviet support, was launched and retook almost all of Eritrea at considerable cost. Thousands of Eritreans were killed and hundreds of thou-sands of refugees fled into the Sudan. But Ethiopia was unable to deliver a *coup de grâce*. The Eritreans clawed their way back into contention and a 'fluid-stalemate' prevailed.

Ethiopia's position was made worse by a revolt in Tigray province, not for independence as in the case of Eritrea but for greater autonomy within Ethiopia. Both Eritrea and Tigray were devastated by the droughts of 1983–5, thousands died of starvation but the wars continued relentlessly. The human suffering was appalling. As the 1980s progressed the war took its toll on Ethiopia, which spent vast sums on the military despite the desperate plight of millions through recurring famine. The government attempted a collectivization of peasant agriculture and tried to resettle up to 1.5 million people in order to over-come the droughts in the north. These were inappropriate, imperialistic, ideo-logical and dictatorial responses to the problems that faced all of the people of Ethiopia and all were unsuccessful. In 1991 Mengistu was overthrown.

In May 1993 the people of Eritrea voted in a referendum for full independence from Ethiopia. Eritrea became Africa's fifty-third sovereign state and at the same time Ethiopia became Africa's fifteenth land-locked state. More importantly Eritrea's independence marked the end of one of the longest running, most destructive wars in post-independence Africa.

38 African imperialism: Morocco and Western Sahara

The Western Sahara problem is essentially similar to that of Eritrea. On ceasing to be a European colony Western Sahara was occupied by a neighbouring African state claiming historic rights. The territory, this time because of its natural resources, is also of considerable value to the occupying power. The people of Western Sahara have been denied the right of self-determination and the guerilla war fought by the Sahrawis against a powerful militarized state with superpower backing has literally run into the sand. Unlike the situation in Eritrea the international community has been involved with the Western Sahara dispute at several levels, namely the International Court of Justice (ICJ), the UN and the OAU. The net effect is painfully slow progress towards a referendum to determine the future of the territory, but that could well solve nothing as accusations have been made that among those eligible to vote will be large numbers of Moroccans who have moved southwards across the border since 1975.

In 1884 Spain claimed the 600 miles (100 km) Saharan coast between Morocco and Mauritania as its 'sphere of influence'. Spanish enthusiasm was limited to the foggy desert coast which faced their valued possession of the Canary Islands. Although France became politically dominant in Morocco from 1911, Spain secured those parts strategically important to it: the northern Rif, facing Spain itself, and the hinterland to Ceuta and Melilla; the southern protectorate opposite the Canary Islands and contiguous with Western Sahara; and the fishing port enclave of Sidi Ifni. Spanish control over the desert interior was minimal. The population was small, mainly nomadic herders whose traditional territories extend beyond the mainly straight-line boundaries. The Spanish census of 1974 put the total population of Western Sahara at 73,500 but the UN estimated more than twice that number. In 1965 deposits of 1700 million tonnes of high quality phosphates were confirmed at Bu Craa. A refinery was built, connected by trunk conveyor to the port of El Aaiún (La'youn). At a stroke Western Sahara was transformed from a desert wasteland into valuable real estate.

After independence in 1956 Morocco was aggressively expansionist, claiming Western Sahara, Mauritania and parts of Algeria on the basis of the sixteenth-century Moroccan empire that had extended as far as Timbuctoo. In 1957 Morocco invaded Western Sahara but was repulsed by Spain. After Algerian independence war flared between Morocco and Algeria in 1963 over the boundary, which the French had failed to define. The conflict centred on

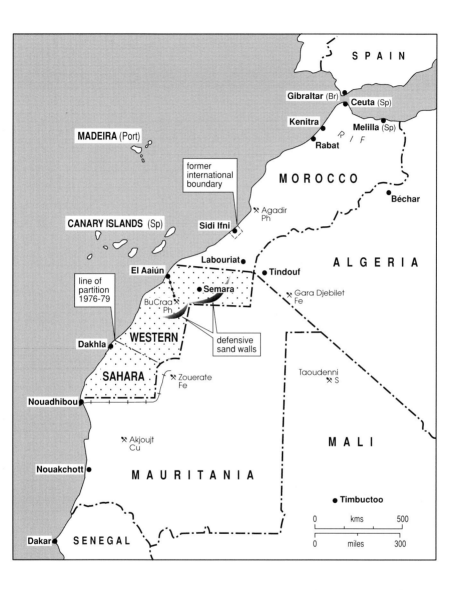

large, unworked iron ore deposits at Tindouf. Morocco was again repulsed. Diplomacy took over: Spain gave up Sidi Ifni in 1968 but left the issue of Western Sahara, Ceuta and Melilla to be resolved. Morocco and Algeria signed a treaty of friendship in 1969, and in the same year Morocco recognized Mauritania, a relinquishing territorial claims on that state. Morocco continued to press for decolonization of Western Sahara, assuming that Spain's withdrawal would be followed by Moroccan rule. But Spain's belated phosphate-led interest resulted in some economic and political devellopment. In 1967 Spain set up the Yema's, an assembly of nominated and elected members to give advice on local administration. In 1973 the Yema's asked that the Sahrawis be allowed self-determination and in 1974, to Morocco's consternation, Spain agreed. Meanwhile in May 1973 a new nationalist movement, the POLISARIO front, had been formed to accelerate political development by direct action.

The advent of the POLISARIO galvanized Morocco into diplomatic action. The UN was persuaded to ask the ICJ to advise on the legal status of Western Sahara before Spanish colonization. The UN also agreed to send a mission to assess the problem on the spot and to visit other interested states. Spain agreed to postpone a referendum in Western Sahara until the UN had received both reports. Late in 1974 Morocco and Mauritania secretly agreed to partition Western Sahara between them when the opportunity arose.

In mid-October 1975 things came to a head dramatically when, within a few days of each other, the UN mission and the ICJ published their findings as Franco, the Spanish dictator, lay dying. Both reports recognized pre-colonial ties between Western Sahara and Morocco and between Western Sahara and Mauritania, but saw no reason to withhold from the Sahrawis the right of self-determination. Spain, playing for time in an awkward interregnum, went back to the UN who suggested a six-month cooling-off period. King Hassan replied on 6 November 1975 by leading 350,000 Moroccans in the 'Green March' across the Saharan border. On 14 November an agreement was signed between Spain, Morocco and Mauritania for Spanish withdrawal in early 1976 and for the partition of Western Sahara between Morocco and Mauritania, Morocco taking the northern two-thirds with the phosphates of Bu Craa.

Ignored in the take-over, the POLISARIO fought on and, in February 1976, the Saharan Arab Democratic Republic (SADR) was set up. With help from Algeria, which gave effective sanctuary, the POLISARIO first chipped away at Mauritania whose vital iron-ore mine at Zouerate and railway to Nouadhibou could not be easily defended, even with French Air Force aid (1976–8), against guerilla attack. The unpopular war, which threatened Mauritania's main resource, drained its fragile economy and built up the army and led inevitably to a military *coup d'état.* The long-serving Ould Daddah government fell, a cease-fire followed and in 1979 Mauritania made peace with the SADR.

The war between the POLISARIO and Morocco settled into stalemate. Morocco held the part that matters economically – the Bu Craa, El Aaiún (La'youn), Semara triangle and pushed out walls of sand hundreds of miles long, topped by sophisticated radar devices to prevent the POLISARIO from continuing their devastating surprise strikes. Behind the wall the Moroccans have found large new deposits of iron ore.

On the diplomatic front the POLISARIO/SADR steadily gained support and recognition within Africa but it was ineffective. A majority of member states were willing to admit the SADR to the OAU at the 1980 Sierra Leone summit, but there was prevarication. The most serious among several incidents was that the OAU failed to meet at Tripoli in August 1982 because more than one-third of member states stayed away.

The OAU was unable to solve the matter and openly split into conservative and radical camps, giving a stunning and damaging display of African disunity. The Western Sahara problem sank into stalemate. Militarily the POLISARIO could only irritate the entrenched Moroccans but they would not go away and, as long as they had a safe base in Algeria, they could not be beaten decisively. Diplomatically the only way forward was through a referendum, as suggested by the OAU in 1981 and later taken up by the UN. But Morocco has refused prior withdrawal and there is seemingly endless wrangling over the vital composition of the electoral roll.

Within Africa Morocco's chief supporters, Senegal, Somalia, Tunisia and Zaire, want the status quo to remain, fearing that a POLISARIO victory would strengthen radical forces in the OAU. Saudi Arabia, the United States and France share these concerns on wider fronts and are prepared to support Morocco with financial aid, military supplies and diplomatic influence. Morocco's Kenitra base has served the United States well as a military staging-post. It is also well recognized that Western Sahara is an issue by which King Hassan is able to deflect criticism of his sometimes precarious and unpopular rule within Morocco. Better the conservative King closely allied with the West than a more radical, even fundamentalist, replacement.

Support for the SADR has primarily come from Algeria. Algeria is bothered by Moroccan expansionism in general but in particular would like to develop the Tindouf iron-ore mines with access to the sea via El Aaiún (La'youn). In April 1983 Algeria and Morocco moved closer together and for the first time in almost eight years opened the boundary between them. In late 1984 Gadafy recognized Morocco's case in Western Sahara in return for Morocco's recognition of the Libyan case in northern Chad. In 1984 the SADR was seated at the OAU with support in particular from Ethiopia, Angola, Mozambique and Zimbabwe. Morocco responded by walking out. But the recognition has led nowhere.

Morocco has extended its succession of sand walls to control almost all of the territory of Western Sahara. In 1993 events were moving slowly towards a UN-endorsed referendum. In prospect the problems surrounding it are enormous, starting with the question 'Who has the right to vote?' Would either party really accept a 'bad' result?

39 Further reading

Asiwaju, A.I. (ed.) (1985) *Partitioned Africans: Ethnic Relations across Africa's International Boundaries*, London: Hurst, and Lagos: University of Lagos Press.

Cervenka, Z. (ed.) (1973) *Land-locked Countries of Africa*, Uppsala: Scandinavian Institute of African Studies.

Firebrace, J. (1984) *Never Kneel Down: Drought, Development and Liberation in Eritrea*, Nottingham: War on Want.

Griffiths, I.Ll. (1986) 'The scramble for Africa: inherited political boundaries', *Geographical Journal* 152 (2): 204–16.

Griffiths, I.Ll. (1989) 'The quest for access to the sea in southern Africa', *Geographical Journal* 155 (3): 378–91.

Hansen, E. (1987) *Africa: Perspectives on Peace and Development*, London: Zed and NJ: United Nations University.

Hargreaves, J.D. (1988) *Decolonization in Africa*, London: Longman.

Hodder-Williams, R. (1984) *An Introduction to the Politics of Tropical Africa*, London: George Allen & Unwin.

Hodges, T. (1984) *The Western Saharans*, London: Minority Rights Group.

Kamil, L. (1987) *Fueling the Fire: U.S. Policy and the Western Sahara Conflict*, Trenton, NJ: Red Sea Press.

Kinnock, G. (1988) *Eritrea: Images of War and Peace*, London: Chatto & Windus.

Laidi, Z. (1990) *The Superpowers and Africa: The Constraints of a Rivalry 1960–1990*, Chicago, Ill: University of Chicago Press.

Manning, P. (1988) *Franco-phone sub-Saharan Africa 1880–1985*, Cambridge: Cambridge University Press.

Ojo, O., Orwa, D.K. and Utete, C.M.B. (1985) *African International Relations*, London: Longman.

Pool, D. (1980) *Eritrea: Africa's Longest War*, Report 3, London: Anti-Slavery Society.

Nkrumah, K. (1963) *Africa Must Unite!*, London: Heinemann.

Selassie, B.H. (1980) *Conflict and Intervention in the Horn of Africa*, New York and London: Monthly Review Press.

Widstrand, C.G. (ed.) (1969) *African Boundary Problems*, Uppsala: Scandinavian Institute of African Studies.

Whiteman, K. (1988) *Chad*, London: Minority Rights Group.

D Economics

40 Poverty

By any material measure, in general the people of Africa are poor. The total GNP of all fifty-two African states in 1990 was only about 7 per cent of that of the United States. Average GNP per caput in Africa is only about one-thirty-fifth of that in the United States. Not only do these figures represent a shocking poverty gap but it is a gap which has grown dramatically wider in the last decade. In 1981 total African GNP was over 12 per cent of that of the United States and average African GNP per caput was one-sixteenth of that of the United States.

These measures of wealth and poverty, though widely used, need to be handled carefully, especially on a comparative basis across cultures and continents and in terms of a single currency. Particular problems include the proper assessment of subsistence agriculture, which is the economic activity of the overwhelming majority of Africans; the inability satisfactorily to account for cultural and environmental factors affecting food, clothing, fuel, and shelter; the use of a monetary common denominator, so that fluctuations in currency exchange rates are sometimes the dominant feature in an index supposed to reflect comparative material well-being; while average GNP per head of population hides possible gross inequalities caused by maldistribution of wealth within individual societies. Nevertheless there is no escaping from the assertions that, in material terms, Africa is a poor continent and that the poverty gap between Africa and industrialized countries, as represented by the United States, is widening rapidly.

There is great contrast in levels of GNP per caput between countries in Africa. At one end of the scale Libya enjoys a GNP per caput of about one-quarter of that of the United States (in 1981 it was one-half), while at the other at least twelve states have a GNP per caput of under US $250 (eleven in 1981). Only nine African states have a GNP per caput of over US $1000 (twelve in 1981). Most of these are well endowed with mineral wealth. Libya and Gabon combine rich oil resources with low populations. Algeria has large quantities of oil and gas with a moderately sized population. South Africa has large mineral resources but the average GNP figure hides the maldistribution of wealth within the state caused by apartheid. The Indian Ocean islands of the Seychelles and Mauritius are comparatively prosperous with tourism and sugar the respective mainstays.

The poorest states in Africa are generally in the Sahel and the Horn, but the poorest of all is Mozambique. In common with Somalia, Uganda, Ethiopia and Chad (no figures are available for Liberia and the Sudan), it is war

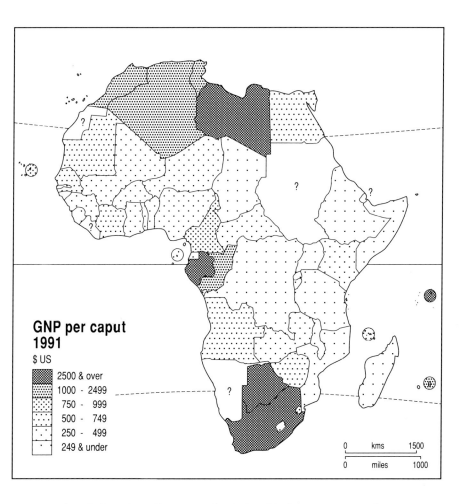

GNP per caput 1991

$ US

2500 & over	
1000 - 2499	
750 - 999	
500 - 749	
250 - 499	
249 & under	

kms 1500

miles 1000

ravaged and there is a direct causal relationship between war and poverty, poverty and war, a vicious, vicious circle of human suffering. The poorest states also have few natural resources and have been hit repeatedly in recent years by natural disasters ranging from crippling drought to devastating cyclones.

Between 1980 and 1991 the growth rate in GNP per caput actually fell in twenty-five African countries out of only forty-seven reporting (fifteen over the decade 1973–82). They ranged from Libya, the richest in terms of GNP per caput, which was hit by falling oil prices, to the war-torn trio of Ethiopia, Somalia and Mozambique (refer to Chapter 69, Statistics).

41 National economies

National economies in Africa are very small. In 1991 South Africa was by far the largest, and only Algeria also had an economy greater than US $50 billion. Egypt and Nigeria had economies of about US $30 billion, Libya and Morocco about US $20 billion and Cameroon and Tunisia just over US $10 billion. Five of the eight largest economies were of Arab North African countries. Meanwhile nine sub-Saharan African states had national economies of less than US $1 billion a year, among them that of St Thomas and Prince Islands, which was calculated at a mere US $42 million in 1991. Here is the economic manifestation of political balkanization.

In a breakdown of Gross Domestic Product (GDP), of the three sectors, agriculture, non-manufacturing industry (mainly mining) and manufacturing, agriculture dominated in twenty-five of the thirty-nine African countries for which statistics were available, and was equal first in one other. In eight countries mining dominated and was equal first in two others. Only in South Africa and Zambia was manufacturing the largest sector, and in Morocco shared equal first place.

In general, in all of the poorest countries the agricultural sector predominated. In Tanzania, Somalia, Mozambique, Burundi and Mali agriculture accounted for over 50 per cent of GDP. Because in most African countries the sector is very poor, to achieve such a high proportion of GDP requires extremely high proportions of people engaged in agriculture, and Africa is very largely a continent of agriculturists. In twenty-one African countries over 75 per cent of the labour force was employed in agriculture and in another sixteen more than 50 per cent. In Rwanda and Burundi 93 per cent of the labour force was in agriculture. There is a strong correlation between a high proportion of GDP derived from agriculture and poverty in Africa.

Non-manufacturing industry (mainly mining, including oil and gas exploitation) accounted for 20 per cent or more of GDP in no fewer than twelve African countries. This represents a doubling of states in this position over the decade, pointing to increased exploitation of mineral raw materials on the one hand (the equivalent of selling off irreplaceable family silver) and the relative decline of other sectors of African national economies on the other. The most mineral-dependent state was Botswana with 53 per cent of its GDP derived from this sector. In addition, all of the big oil producers, Libya, Nigeria and Algeria, along with Gabon (oil and manganese), Guinea (aluminium) and Namibia (diamonds and uranium), had mining sectors which accounted for 30 per cent or more of GDP. All of the richest countries in

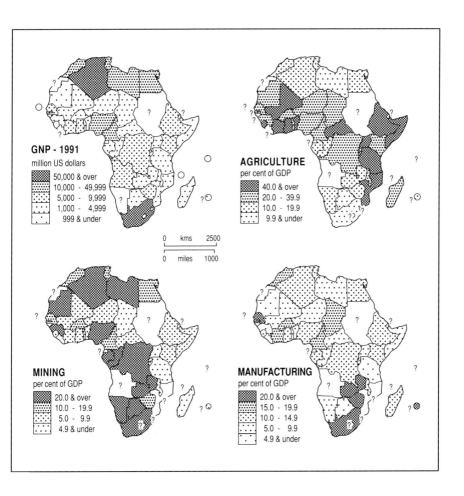

GNP - 1991
million US dollars
- 50,000 & over
- 10,000 - 49,999
- 5,000 - 9,999
- 1,000 - 4,999
- 999 & under

0 kms 2500

0 miles 1000

AGRICULTURE
per cent of GDP
- 40.0 & over
- 20.0 - 39.9
- 10.0 - 19.9
- 9.9 & under

MINING
per cent of GDP
- 20.0 & over
- 10.0 - 19.9
- 5.0 - 9.9
- 4.9 & under

MANUFACTURING
per cent of GDP
- 20.0 & over
- 15.0 - 19.9
- 10.0 - 14.9
- 5.0 - 9.9
- 4.9 & under

Africa had large mining sectors, confirming the importance of mineral development to accelerated development in Africa.

Manufacturing is the least developed of the three sectors in African countries. Only South Africa, Zimbabwe, Zambia, Mauritius and Senegal derived more than 20 per cent of GDP from manufacturing. These states have long had government policies encouraging industrialization as a vital part of development strategy. But, for the most part, African countries depend heavily on the industrialized world for imports of manufactured products and increasingly lose out as the terms of trade consistently move in favour of manufactures and against raw materials.

42 Traditional economic systems

The vast majority of Africans are rural dwellers living directly off the land by different subsistence systems which are adapted to their various physical environments. Some small groups of hunter-gatherers still survive in the remoter parts of Africa, notably the San (Bushmen) in the 'lost world of the Kalahari', and Pygmy groups in the tropical rain-forest areas. They live by hunting animals, collecting edible fruits and roots, and sometimes tending small 'gardens' which are visited on a seasonal cycle as part of a nomadic round within a well defined territory. If given adequate space by encroaching pastoralists and cultivators, hunting and gathering is a viable means of subsistence which calls for a deep knowledge of, and an intimate and infinitely complex relationship with, the physical environment.

Various forms of pastoralism prevail in the drier parts of Africa. Some are nomadic, some settled and some, as in Botswana, a combination of both. All are extensive in their use of land and, in different ways, represent complex human responses to the problems of living in difficult semi-arid environments. Nomadic pastoralists migrate with their animals in search of grazing and water. Their wanderings follow well-defined annual circuits from one known source of food and water to another, returning each year to a home base. The systems are vulnerable to the vagaries of climate in marginal lands where rainfall is low, notoriously unreliable and subject to periodic drought.

In large parts of tropical Africa permanent cultivation is not possible because of rapid soil impoverishment once the natural vegetation cover has been removed. Shifting cultivation is a response to these environmental conditions. A plot of land is carefully selected, cleared of trees and bush and the accumulated debris burnt to produce ash. The plot is then planted, perhaps for two years, before the cultivator moves to a new plot, leaving the forest to regenerate and replenish the old plot. The many forms of shifting cultivation are all extensive in their use of land and can only be practised in areas of low population density. As the population of Africa grows and with it pressure on the land, so shifting cultivation contracts.

Permanent cultivation is possible only in relatively few favoured areas in Africa: on rich volcanic soils, as in Rwanda and Burundi, and in river valleys, such as the lower Nile and inland Niger delta, where fertility is renewed annually by silt deposits from seasonal floods. In a few areas, such as Buganda, it is possible to grow perennial crops, plantains and bananas, which replenish the soil in a natural cycle. Permanent cultivation usually supports high densities of population.

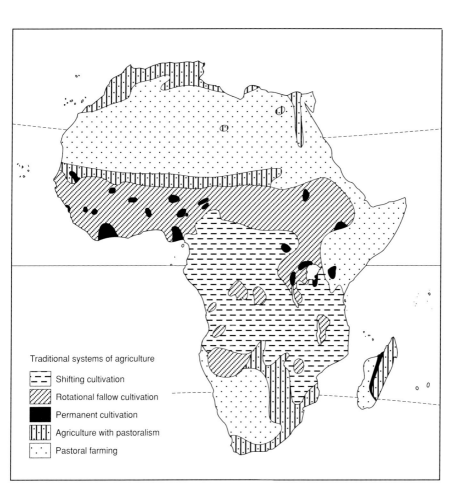

Traditional systems of agriculture

- Shifting cultivation
- Rotational fallow cultivation
- Permanent cultivation
- Agriculture with pastoralism
- Pastoral farming

Traditional economic systems are under great pressure throughout Africa. Balances between humans and the environment have been upset in the cause of progress elsewhere. The dam that prevents flooding also prevents soil renewal, the wage-earner enticed to the mines is one less hand in the fields, the infant's life saved is one more mouth to feed. Too little effort has been made to understand the traditional economies of Africa, in particular their harmony with the physical environments in which they thrive. Until that is done they cannot properly be helped to accommodate the high speed of necessary change.

43 Causes of famine

Dawn, and as the sun breaks through the piercing chill of night on the plain outside Korem, it lights up a Biblical famine, now, in the twentieth century. This place, say workers here, is the closest thing to hell on earth. Thousands of wasted people are coming here for help. Many find only death. They flood in every day from villages hundreds of miles away, felled by hunger, driven beyond the point of desperation. Death is all around.

Michael Buerk's introduction to the eight-minute news item from Korem in Ethiopia on BBC-TV in October 1984 was almost as graphic as the news film itself. For many, including highly placed UN officials, it was their first knowledge of that particular crisis in Africa.

Famine and starvation are not new to Africa. In a 'normal' year as many as 100 million Africans are malnourished. Africa suffers many separate famine crises of different place, time, cause and effect. Since 1971 many countries in Africa have experienced famine: Mauritania, Mali, Burkina Faso, Niger, Chad, Sudan, Uganda, Ethiopia, Somalia, Angola, Zambia, Botswana, Zimbabwe and Mozambique; but not at the same time, or to the same degree, or from the same causes.

Natural disasters, such as droughts, typhoons or plagues of locusts, often do trigger famine in Africa but it would be wrong to assume that physical factors are alone responsible. Their initiating, causal role must be recognized but equally it must be seen that a progression from natural disaster to famine to death by starvation is not inevitable.

In many countries the growth in total GNP is more than offset by population growth, so that GNP per caput declines. In rural Gambia, for example, it is a delight to be greeted in a rural village by a surge of young children, only to ponder later how they, let alone future generations, will be fed. Population growth rates in many parts of Africa give cause for concern but it would be wrong to consider population in isolation or to think it the prime cause of famine.

The colonially induced switch to cash crops for export from food crops for home consumption has led to many African countries being dependent on food imports at unaffordable prices on world markets. Recruitment of rural Africans to work in mines has had a great impact on African agriculture. The attraction of towns has impoverished the rural areas. Africa has been subjected to a century or more of exploitation and systematic underdevelopment from outside. Within African countries urban élites have done the same

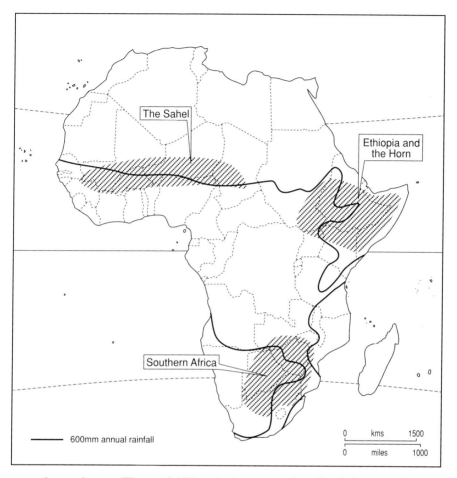

The Sahel

Ethiopia and the Horn

Southern Africa

—— 600mm annual rainfall

| 0 | kms | 1500 |
| 0 | miles | 1000 |

to the rural areas. The rural African is the most exploited and the most likely to die of starvation.

Governmental interference, in the way of taxes, fixed prices, marketing boards and production quotas, also contributes to crisis in rural Africa. It keeps urban food prices low, encourages rural/urban migration and favours cash crops and food imports, often of exotics, for the urban élites.

Civil wars, too, have been rampant over much of the last twenty years in most of the states that have experienced famine. The wars are often the product of political immaturity, another part of Africa's colonial inheritance. People in war-torn areas cannot produce food but can only consume it after they flee; suffering horribly in crowded, unhygenic refugee camps, they can but hope that food aid will find a way to them through the war zone.

44 Cash crops and colonialism

Cash crops, as the name implies, are grown to be sold, in the African context usually on export markets to earn foreign exchange. They are the mainstay of many African economies, and in some states are the only exports. They are part of a system of agriculture and commodity trading which was devised largely in the colonial period. The crops grown are tropical or sub-tropical and provide a range of exotic commodities for the metropolitan countries that controlled the colonial empires in Africa. The cash earned enabled the colonies to pay their way.

To help to guarantee supplies, the system of tropical plantation agriculture grew up. In territories where physical conditions were suitable, land was taken and planted with crops, usually on the basis of monoculture. For some crops, such as rubber, considerable investment was called for, not just in clearing the land and planting but also in waiting for the trees to mature. Terrible mistakes were made as a result of selecting the wrong place or underestimating the task of clearing the land or the incidence of disease. Perhaps the most spectacular failure was the 'groundnuts affair' in post-war Tanganyika. The British Colonial Office, in an apparently attractive project, among other things tried to use ex-Second World War Sherman tanks to clear and deracinate vast areas around Dodoma, which turned out to be totally unsuited for groundnut monoculture. Generally, however, the plantation system worked and, when independence came, it was maintained by the newly independent countries which thirsted after foreign exchange to help to fund economic development.

It is only in recent years that the full implications of implanting this system of agriculture have been fully scrutinized, especially as countries dependent on cash crops have found that the terms of trade have moved consistently against them, making cash crops a far less attractive prospect than they once seemed.

Cash crop production required, in addition to capital investment, technical and managerial skills from the metropolitan country and two important local commodities, land and labour. Almost invariably land taken for cash crops was land alienated from traditional occupance. This not only caused trouble at the time of alienation and disrupted traditional agriculture but also, as populations grew in the colonial and post-colonial periods, was a root cause of land shortage. Such shortage was particularly acute in some countries, notably Kenya and Zimbabwe, where a great deal of land was alienated, much of it for cash crops in one form or other. Peasants, who were forced to enter the money economy as a result of the device of imposing a hut or poll tax, provided

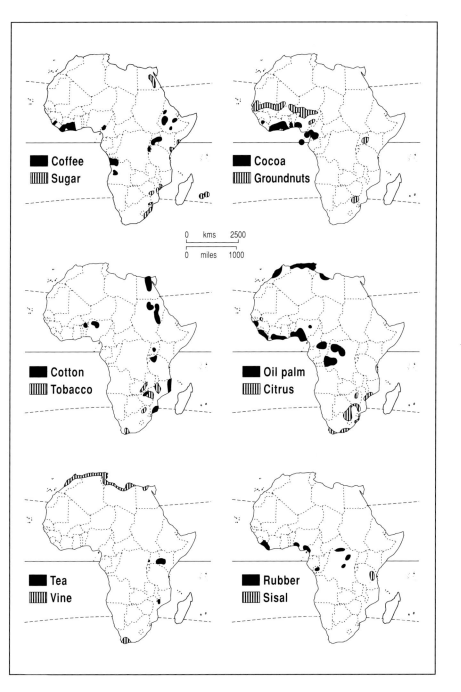

Coffee
Sugar

Cocoa
Groundnuts

0   kms   2500
0   miles   1000

Cotton
Tobacco

Oil palm
Citrus

Tea
Vine

Rubber
Sisal

labour for the plantations and so helped to deplete the labour supply to the traditional sector, without any compensatory input. To make matters worse, the plantations took only the best workers, young and healthy. In modern Africa emphasis on cash crops is often carried to extremes, to the extent that production of food crops for home consumption is insufficient and so people go hungry or staple foods are expensively imported at high world market prices.

Monoculture is peculiarly vulnerable to disease and adverse weather conditions, and to economic problems such as product substitution and price fluctuation. Countries are sometimes almost totally dependent on a single cash crop commodity for all of their foreign exchange earnings. The prosperity of Mauritius depends almost entirely on cane sugar, a commodity which is subject worldwide to overproduction and low prices, not least because many European countries have become self-sufficient through growing beet-sugar. Senegal and the Gambia are both heavily dependent on groundnuts which are exported whole, shelled or as oil or cake. Egypt, Sudan, Mali and Chad are heavily dependent on cotton, another commodity which has suffered from competition from substitutes in the form of man-made fibres. Commodity prices, which are usually determined in industrialized countries, are subject to sharp fluctuations resulting from many factors, ranging from crop failure in one part of the world to stockpiling in another. Attempts to set up international commodity agreements, which guarantee basic prices and impose production quotas, do not extend to all commodities and do not always work. Commodity prices have fallen, relative to prices of manufactured goods, and African countries dependent on cash crops suffer accordingly. The main problem is overproduction of various commodities which are grown under a system that is well dispersed throughout the former colonial empires of the world.

The cash crop sector, while earning foreign exchange through its exports, is also a high foreign exchange spender because it utilizes a relatively high level of technology which has to be imported. Little foreign exchange is left to be used elsewhere in the economy, perhaps to fund diversification.

Most cash crops are exported as raw materials. African countries do little processing themselves before export, not least because of resistance from the industrialized consumer countries which want to enjoy the advantages deriving from manufacture. Africa produces about 7 per cent of the world's cotton but accounts for about 12 per cent of the world's exports. Africa produces under 3 per cent of the world's cotton yarn and cotton fabric, and accounts for only 2.4 per cent of exports of cotton fabric.

Some African countries, while still dependent on cash crops, have instigated a move away from the plantation system. Peasant farmers have been

Cash crop production, 1990 (1000 tonnes)

Cash crop	World	Africa(%)	Leading African producers (%)	
Cocoa	2398	1263 (52.7)	Ivory Coast	700 (55.4)
			Ghana	245 (19.4)
			Nigeria	155 (12.3)
Dates	3431	1268 (37.0)	Egypt	580 (45.7)
			Algeria	212 (16.7)
			Sudan	130 (10.3)
Sisal	378	102 (27.0)	Kenya	39 (38.2)
			Tanzania	30 (29.4)
			Madagascar	21 (20.6)
Cashew nuts	479	107 (22.3)	Mozambique	49 (45.8)
			Tanzania	20 (18.7)
			Kenya	12 (11.2)
Groundnuts (in shell)	23109	4765 (20.6)	Nigeria	1166 (24.5)
			Senegal	698 (14.6)
			Zaire	430 (9.0)
Coffee	5964	1204 (20.2)	Ivory Coast	219 (18.2)
			Ethiopia	195 (16.2)
			Uganda	168 (14.0)
Peppers	9083	1690 (18.6)	Nigeria	800 (47.3)
			Egypt	245 (14.5)
			Algeria	210 (12.4)
Palm oil	11084	1762 (15.9)	Nigeria	900 (51.1)
			Ivory Coast	214 (12.1)
			Zaire	180 (10.2)
Bananas	45845	6210 (13.5)	Burundi	1608 (25.9)
			Tanzania	1380 (22.2)
			Uganda	490 (7.9)
Tea	2522	323 (12.8)	Kenya	197 (61.0)
			Malawi	39 (12.1)
			Tanzania	20 (6.2)
Pineapples	9652	1191 (12.3)	South Africa	265 (22.3)
			Kenya	202 (17.0)
			Zaire	143 (12.0)
Avocados	1463	169 (11.6)	Zaire	45 (26.6)
			Cameroon	35 (20.7)
			South Africa	35 (20.7)
Aubergines	5761	574 (10.0)	Egypt	400 (69.7)
			Sudan	80 (13.9)
			Morocco	32 (5.6)

Cash crop	World	Africa(%)	Leading African producers (%)	
Cotton lint	18457	1317 (7.1)	Egypt	330 (25.1)
			Sudan	125 (9.5)
			Ivory Coast	108 (8.2)
Sugar cane	1035096	72982 (7.1)	South Africa	18700 (25.6)
			Egypt	11143 (15.3)
			Mauritius	5548 (7.6)
Rubber	5108	286 (5.6)	Nigeria	80 (28.0)
			Ivory Coast	74 (25.9)
			Liberia	70 (24.5)
Tobacco	6634	367 (5.5)	Zimbabwe	139 (37.9)
			Malawi	91 (24.8)
			South Africa	34 (9.3)
Wines	29210	1134 (3.9)	South Africa	945 (83.3)
			Algeria	100 (8.8)
			Morocco	50 (4.4)

Source: FAO (1991) *Production Yearbook 1990.*

encouraged to grow cash crops interplanted with subsistence crops. To do this successfully it is necessary to have both an efficient growers' co-operative and a marketing board-type of system which offers guaranteed prices, makes seed available, provides technical assistance and ensures that crops are efficiently collected, graded and exported. On the other hand, this form of organization is seen by the IMF as unwarranted government interference which in large part is responsible for the subjugation of the rural sector to the priorities of the urban sector; for example, by keeping food prices low and syphoning off too many of the rewards of agricultural enterprise for other uses. With the active encouragement of the IMF, there has been a movement of private urban capital into specialized, small-scale but high cash-yield, cash crop production. These enterprises grow vegetables, such as aubergines and peppers, to be air-freighted to the European market.

Although differently organized, this is really more of the same. Cash crops produced in Africa give most benefit to people outside Africa, on the one hand providing an enormous range of exotic fruits and vegetables for the shelves of northern supermarkets and, on the other, giving northern processors and manufacturers the opportunity to profit on the backs of the African producers. For the producer the whole cash crop business is fraught with difficulty; there are problems not only of production but also of marketing, such as freighting, dependence on sales agents in industrialized countries and, above all, the wide fluctuations in price which are beyond the producer's control. A more radical solution is necessary to ease the lot of the many African countries that are caught up in cash crops as part of a world trading system which works patently to their disadvantage.

45 War, disaster and refugees

Many of the poorest states of Africa are so because they have been buffeted by human and natural disasters. Millions of people in Africa have been displaced, some within their own countries, others to become international refugees. Within Africa there is a strong correlation between war and poverty, famine and death through starvation.

GNP per caput, military expenditure and arms imports for selected African countries

State	GNP per caput (US $)	Military expenditure (per cent GNP)	Arms imports (million US $)	Periods of war
Angola	620	21.5	3592	from 1975
Chad	220	3.8	100	from 1978
Ethiopia	120	13.6	629	from 1961
Mozambique	70	10.4	19	from 1977
Somalia	150	3.0	43	from 1960
Sudan	–	5.9	189	1956–72 and from 1982
Uganda	160	4.2	33	from 1979

Sources: IBRD (1992) *World Bank Atlas 1992* and UNDP (1992) *Human Development Report 1992.*

Wars disrupt food production. Fields cannot be tilled, seeds cannot be planted, seed stock is plundered. Crops cannot be harvested, harvests and livestock are commandeered. Peasants flee in terror before rampaging armies to 'food distribution points', to insanitary, unhealthy 'refugee camps'; those who survive are later often forcibly resettled against their will. Relief work is highly dangerous and is hampered by damaged infrastructure. Scarce resources and even short-term famine relief is diverted to the military.

The wars of Africa have many causes. Somalia refused to accept its colonial boundaries at independence because they excluded a third of Somali speakers. When not engaged in irredentist war the country is wracked by bitter civil war. Eritrea was also at war with Ethiopia, in a war of African imperialism, aided and abetted by superpower geo-political manoeuvering. In Chad the flames of intermittent civil war are fanned by the imperialism of Libya which covets the Aouzou strip. In Uganda the civil war is dying down after more than a decade of slaughter. In Sudan there is a constant rekindling of civil war between Muslim north and non-Muslim south. The long-running civil wars in Angola and Mozambique were part of the death throes of

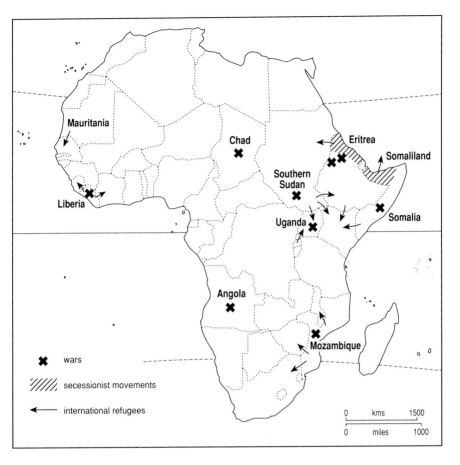

Key (on map):

✖ wars

▨ secessionist movements

← international refugees

	0	kms	1500
	0	miles	1000

apartheid. South African destabilization and sponsorship of anti-government rebel groups have devastated large areas of both countries. In the mid-1980s the South Africans and their allies deliberately employed famine creation in Mozambique as a weapon in defence of the apartheid state. In Angola the world superpowers fought out their rivalries by proxy. (Refer to Chapter 70, statistics.)

Wars breed refugees who spread the effects of war into nearby states: Ethiopians into Sudan and Kenya; Sudanese into Ethiopia, Kenya and Uganda; Somalis to Kenya and, as boat people, across the Gulf of Aden to Yemen; Chadians to Sudan; Mozambicans to Malawi, Zimbabwe and, under the electrified fence, into South Africa. Millions are displaced, to die of starvation and disease, those who survive are deprived of any means of earning a living and their condition is a root cause of poverty in Africa.

46 Development and population growth

In 1799 the Rev. T.R. Malthus argued that food resources growing in arithmetical progression would limit the growth of population which tended to grow in geometrical progression. He addressed the British experience in the industrial revolution and his dire warnings proved ill-founded. Nevertheless, in the late twentieth century, his work has been dusted off amid assertions that the relationship between population growth and food resources has again become of critical importance, particularly in Africa where millions have died of starvation and many more are threatened by it. 'Neo-Malthusian' analyses of the African population growth/inadequate food supply/starvation syndrome have been put forward, asserting that a major cause common to many of the crises in Africa is simply too many mouths to feed.

Visits to rural Africa subjectively confirm that there is a prima-facie case to answer. On entering villages in remotest Gambia and Senegal, one's vehicle is immediately surrounded by crowds of beaming children and their young mothers who are usually nursing small babies. The scenes are delightful, particularly as there is no sign of malnourishment, but are cause for later sombre reflection. Here in the flesh is the wide-based population pyramid. Fifteen-year-old girls in these small Muslim rural communities marry, soon bear children of their own and go on doing so, year in and year out, until they die. Life expectancy is about 43 years and many females die younger from causes often associated with childbirth. A sobering thought when looking at the crowds of youngsters is that in just fifteen years they may be the parents of a new, much larger generation.

Between 1960 and 1990 the annual rate of population growth in sub-Saharan Africa was 2.8 per cent whilst the forecast for 1990–2000 is 3.2 per cent (which would double the population in just twenty-two years). These compare with world averages of 1.8 and 1.7 per cent and industrial countries' averages of 0.8 and 0.5 per cent. Population estimates for Africa are: 1960 265 million, 1990 615 million and 2000 835 million. Thirty-two (of fifty) African countries are forecast to have annual population growth rates of 3.0 per cent or over during 1990–2000, compared with twenty-nine countries in 1980–90. In eighteen African states annual population growth rates for 1990–2000 are forecast to be lower than in 1980–90. High fertility rates (6.0 and over) prevail in most African countries (industrial countries are under 2.0), and fell in only twenty-four countries between 1970 and 1990. Crude death rates are also very high in many African states, with four over 20.0 and only ten under 10.0. There

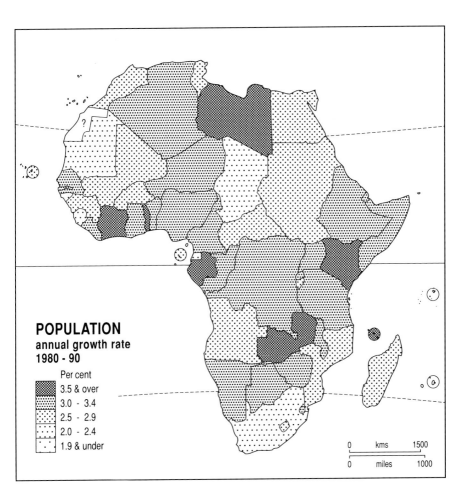

POPULATION
annual growth rate
1980 - 90

Per cent
- 3.5 & over
- 3.0 - 3.4
- 2.5 - 2.9
- 2.0 - 2.4
- 1.9 & under

is ample scope for lowering death rates and that will contribute to further population increase. In no area is improvement more needed than in child care: in 1989, in sub-Saharan Africa, infant mortality averaged 108 per 1000 live births and under-five mortality 179 per 1000 live births. In four African states infant mortality was over 150 and in six under-five mortality was over 250.

Families are large because of the labour value of children, the lack of social security safety nets and, ironically, high infant mortality. Fundamental perceptions need to change. Education and birth control are essential elements in any development strategy but it is wrong to consider population in isolation or to think it the prime factor in Africa's problems (refer to Chapters 9, Population, and 72, Population statistics).

47 International development aid

Many African countries are dependent on large amounts of Official Development Assistance (ODA) from industrial countries. Africa as a whole received over US $16,000 million in ODA in 1989, nine African countries each received more than US $500 million, with Egypt alone receiving US $1578 million. In proportion of total GNP the largest recipients are the poorest countries of Africa. The Gambia received ODA equivalent to over half of its GNP, Mozambique and Somalia over 40 per cent. Seven other African states received ODA equivalent to over 20 per cent of their GNP and more than half of the independent states of Africa over 10 per cent.

Such a degree of dependence is alarming, even if ODA were solely given by 'a generous feeling that wealthier countries have a moral responsibility to help the poor'. The reality is that the motives for ODA are often less altruistic, for example, related to perceived global military strategies. Much ODA is in the form of costly and destructive armaments which have more to do with maintaining in power a regime acceptable to the donor than with alleviating poverty. Flow of ODA is very largely determined by politics rather than need. ODA is sometimes a vehicle of neo-colonialism, used to seek opportunities to develop export markets for goods from the industrial country donor. It is usually earmarked by the donor for a particular use or project and the recipient countries are not free agents in spending the incoming money. ODA frequently has strings attached, that is, the money is given conditionally upon the recipient country following a particular course of action which is dictated, either overtly or covertly, by the donor.

Even where donor countries are prepared to make constructive contributions they prefer to assist finite projects rather than become involved in programmes with a large element of recurrent expenditure. There is a very much better chance of a large dam being built with ODA assistance than there is of the same amount of money being given to education or family planning which is often viewed as a bottomless pit of ongoing commitment. As a result too little ODA is used in supporting areas of social expenditure.

Industrial countries have a poor record as ODA donors. In 1989 only the Nordic countries and the Netherlands (Norway 1.04 per cent of GNP, Sweden 0.97, Denmark and the Netherlands 0.94) had levels of assistance above the UN-set standard. The United States (0.15), despite its huge economy, occupied only second place to Japan (0.32) in total amount of ODA given. Total United Kingdom ODA was less than half of that of France and, along with the United States, its contribution as a proportion of GNP fell over the

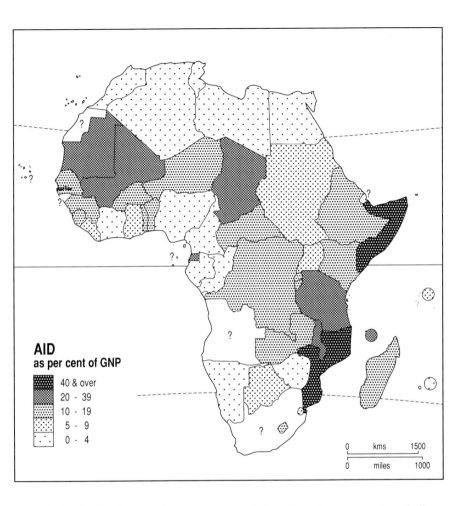

AID
as per cent of GNP

▨	40 & over
▦	20 - 39
▤	10 - 19
⋮	5 - 9
∴	0 - 4

period 1960–89 (from 0.56 to 0.31 per cent). In 1989 only one-quarter of all ODA went to LDCs, and since then much has been directed away from, for example, Africa, to eastern Europe and the former Soviet Union. About 37 per cent of Nordic ODA went to LDCs compared with 19 per cent of that from North America.

Non-governmental organizations (NGOs) make a vital contribution. Over 2000 NGOs contribute the equivalent of less than 10 per cent of ODA but their work is better targeted at LDCs and they do much to mobilize voluntary public support in industrial countries for help to places such as Africa (refer to Chapter 70, Statistics).

48 Debt, the IMF and restructuring

One of the features of Africa through the 1980s was the way in which so many countries fell heavily into debt. By 1988 in thirteen African countries the national debt burden was greater than annual GNP. Debts had to be rescheduled, often through new loans from the International Monetary Fund (IMF), who then used their financial might to impose their form of economic orthodoxy on the debtor country by requiring a restructuring of the economy. The IMF, the World Bank and those industrial countries that are owed money by African states have gone further, using their financial leverage to insist on political reforms including, in some cases, a return to civilian rule and multi-party democracy. About half of all African states, varying in size from the Gambia to Nigeria and in distance from Algeria to Lesotho, have agreements with the IMF so that the impact of IMF policies in Africa is widespread.

The poorest countries (and the poorest people therein) are often the most seriously affected by Africa's debt crisis. They were granted loans, often at high interest rates, by the governments and private banks of industrial countries and then through force of circumstance have been unable to keep up repayments. In the 1980s international credit was made available too easily, with an almost criminally inadequate regard to ability to repay which paralleled the easy availability of domestic credit in industrial countries. High interest rates caused the debt burden to increase, sometimes to a point where it was unrealistic. Mozambique, devastated by civil war, South African destabilization, cyclones and drought, has debts amounting to 376 per cent of GNP. The disasters that Mozambique has faced not only impair the ability to repay loans but also have destroyed much of what the loans were spent on in the first place. Whilst Mozambique is the extreme case and Somalia (debt 185 per cent of GNP) has shot itself (repeatedly) in the foot, countries such as Congo, Mauritania, Madagascar, Tanzania, Egypt, Gambia, Zaire, Zambia, Mali, Equatorial Guinea and Nigeria can offer no such dramatic explanation for owing more than their annual GNP. For some it was simply poor 'house-keeping', others found their ability to repay loans hit by population increase, resource exhaustion or steeply declining terms of trade in export commodities.

In November 1991 a meeting of Western donor governments and aid agencies met in Paris and suspended approval of new aid to Kenya subject to reviewing its record on human rights, corruption and multi-party democracy. Aid has been withheld from Malawi and Zaire. In several Francophone African states 'national conferences' have been held to challenge dictatorial military rule. Elections are breaking out all over Africa. The least bad are

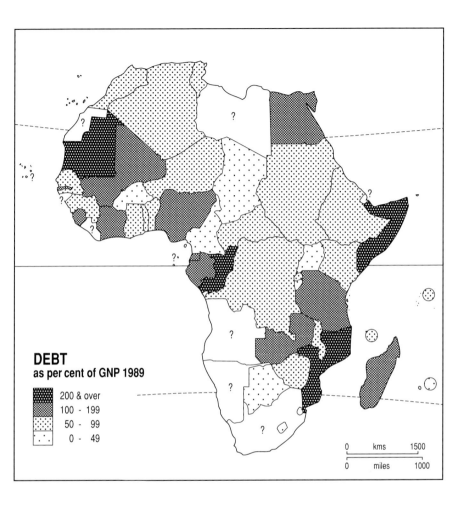

DEBT
as per cent of GNP 1989

- 200 & over
- 100 - 199
- 50 - 99
- 0 - 49

| 0 | kms | 1500 |
| 0 | miles | 1000 |

often the first to go, for example, Kenneth Kaunda in Zambia in elections in
1991, whilst the worst hang on. Is it not conceivable that one-party rule is
appropriate for an African country, which is perpetually facing economic
crisis, in the same way as it was for Britain facing crisis in 1940? Whilst many
economic aims of the IMF are reasonable, is not the thirst for privatization
and market forces simply the pursuit of an economic ideology that is not at all
suited to African states at the bottom of an economic system which is stacked
against them? Should Africa be urged to produce, often at the expense of local
food production, cash crops which are subject to price crashes and mainly for
the benefit of others? (refer to Chapter 70, Statistics).

49 Ghana and Ivory Coast: different development paths

Ghana and Ivory Coast are neighbouring states. Ghana has the larger population (14.9 million to 12.2 million) whilst Ivory Coast has the greater area (322 thousand sq. km to 239 thousand sq. km). In Ghana 49 per cent of the labour force is employed in agriculture, in Ivory Coast it is 46 per cent. Non-manufacturing industry (mainly mining) employs 7 per cent in each state. Ivory Coast's manufacturing sector is larger at 17 per cent of the labour force compared with Ghana's 10 per cent, but Ghana's service sector is larger at 34 per cent compared with 30 per cent. There are no startling differences.

However, at US $8.9 billion the GNP of Ivory Coast is significantly larger than that of Ghana at US $5.8 billion. Because of the greater population of Ghana the GNP per caput of Ivory Coast at US $730 is almost twice that of Ghana at US $390. Why are there these differences when both states have similar economies which are largely dependent on cash crop exports?

Ghana, having set the pace to independence in black Africa, also tried to break the economic mould. In Kwame Nkrumah it had probably the best and most radical leader of the new Africa, who used a healthy economic base, built on cash crops, for centrally planned development founded on industrialization. Infrastructure was built in the form of roads, a new port and the Akosombo hydroelectric dam on the Volta river. The dam was part of an integrated plan for mining Ghanaian bauxite and using the cheap electricity for an alumina plant and smelter. Nkrumah, in an uphill fight with American industrialists, was forced to settle for the dam and a smelter which was to be fed with alumina from the West Indies. The plans were overambitious, infrastructural investment could not yield immediate returns and money was wasted on prestige projects. The economy became overstretched and in 1966 Nkrumah was overthrown. For all his vision, ability, drive and a good launching platform, even Nkrumah was unable to beat the system. Politically unstable, Ghana then had a succession of governments as the economy declined.

Ivory Coast followed different political and economic paths. Felix Houphouët-Boigny has been in power since independence and has made Ivory Coast a client state of France. About 50,000 French are resident in Ivory Coast, working in the civil service and professions. Economic priority is given to cash crop production, namely coffee, cocoa and palm oil, in conformity with the world economic system. The strategy was seen to pay off in the comparative GNP per caput figures.

But the HDI of both states is 0.311. Ghana's rank is higher than that of its GNP per caput, Ivory Coast's is 21 places lower. Life expectancy in Ghana is 55 years, in Ivory Coast 53 years. Adult literacy rate in Ghana is 53 per cent, in Ivory Coast 49 per cent. The final twist is in other economic statistics. Ghana's debt is 44 per cent of GNP, Ivory Coast's 93 per cent. Ghana's GNP per caput fell by 0.6 per cent per annum in 1980–90, Ivory Coast's by 3.7 per cent. Ghana's de-militarized ruler, Jerry Rawlings, runs a tight ship and plans a return to multi-party democracy in 1993 whilst Houphouët-Boigny takes his turn at wasteful prestige projects such as the vast cathedral at his home town of Yamoussoukro. Neither state can be satisfied with its cash crop producer role in the world economic system; Nkrumah's aim was right!

50 Minerals and mining

Africa is extremely rich in minerals and so mining is one of the few means available to African countries of accelerating economic development. The richest states (in terms of GNP per caput) of mainland Africa, namely Libya, Gabon, South Africa, Algeria and Botswana, are *all* dependent for their prosperity on mineral exploitation and many other states in Africa would like to emulate them. But there are serious drawbacks: minerals are a non-renewable resource, demand for individual minerals fluctuates and with it prices and proceeds from sales. Minerals are mainly exported as raw materials, so that most of the benefit goes to others. For a mineral-rich country it is at best a race to use the wealth generated by mining to diversify and modernize the economy and put it on a sound footing where growth is sustainable before the minerals run out. At worst mineral wealth is squandered in a once-and-for-all bonanza followed by economic disaster.

Africa's share of world mineral resources is impressive, though production and share in world markets have declined over the last decade. Partly this is because particular reserves have been exhausted, partly it is because Africa seems to have received less investment capital, perhaps put off by political instability in many regions. Other areas of the world have gone ahead, somewhat at Africa's expense. Nevertheless in 1990 Africa accounted for over 50 per cent of world production of the strategic minerals of cobalt, platinum and vanadium; about 25 per cent or more of diamonds, chromium, manganese and phosphates; 10 per cent of the world production of that resource so critical to modern development, petroleum; and more than 5 per cent of ten other leading minerals.

Through the 1980s Nigeria, Libya, Algeria and Egypt were Africa's largest producers of petroleum; in 1981 they accounted for 87 per cent of African output. In 1990, however, they together produced 72 per cent. Other African producers included Angola, Gabon, Congo and Tunisia, all of which were producers in 1981, and Cameroon which is a new producer. The small production of these countries was significant for them in two ways. First, they did not have to spend large sums of scarce foreign exchange on oil imports and, second, the modest quantity of oil exports for the first three, which was large in relation to their small overall economies, represented over 80 per cent of their total exports.

Petroleum apart, the greatest concentration of mineral wealth is in southern Africa, the world's largest single source of gold, diamonds, cobalt, platinum, chrome and copper. South African gold comes from an arc of reefs

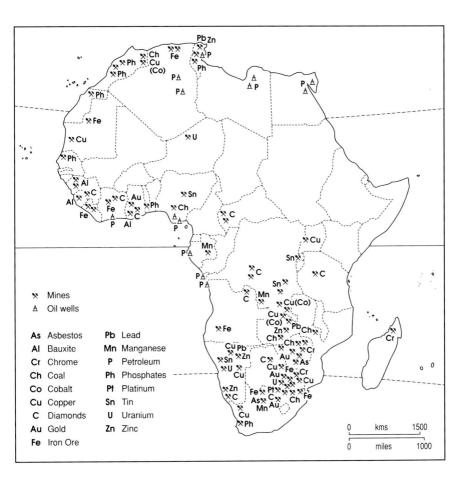

	Mines
⚒	Mines
⚓	Oil wells

As	Asbestos	**Pb**	Lead
Al	Bauxite	**Mn**	Manganese
Cr	Chrome	**P**	Petroleum
Ch	Coal	**Ph**	Phosphates
Co	Cobalt	**Pt**	Platinum
Cu	Copper	**Sn**	Tin
C	Diamonds	**U**	Uranium
Au	Gold	**Zn**	Zinc
Fe	Iron Ore		

	0	kms	1500
	0	miles	1000

300 miles (480 km) long in the southern Transvaal and the Orange Free State. In 1990 South Africa produced 610 tonnes of gold. Kasai province in Zaire is the world's largest source of industrial diamonds. In Botswana, after independence, diamond pipes were discovered under the Kalahari sands, and new discoveries have enabled South Africa to maintain major production which goes back to the Kimberley diamonds discovered in 1870. In Angola the Lucapa region in the north-east has deposits similar to those of Zaire, whilst the sea beaches of southern Namibia have yielded consistently good quality gemstones. The Bushveld Complex, around Rustenburg north-west of Johannesburg in the Transvaal, holds about 75 per cent of world reserves of platinum group metals and has another rare metal with 'high-tech' uses, chromium, which is also found in Zimbabwe. The copperbelt, bisected by the

Africa's share of world production of selected minerals, 1990

Mineral	Africa per cent of world production	Main African producer countries per cent of African production
Cobalt ore	75.9	Zaire 76, Zambia 20.
Vanadium	51.3	S. Africa 100.
Diamonds	49.7	Zaire 45, Botswana 32, S. Africa 16.
Chromium ore	40.7	S. Africa 86, Zimbabwe 12.
Gold	36.9	S. Africa 94, Ghana 3.
Manganese ore	27.0	S. Africa 59, Gabon 34.
Phosphates	24.7	Morocco 56, Tunisia 17, S. Africa 8.
Bauxite (alumina)	18.3	Guinea 90, Sierra Leone 8.
Uranium	14.2	Namibia 35, Niger 30, S. Africa 27.
Copper ore	12.6	Zambia 43, Zaire 38, S. Africa 17.
Mercury	12.6	Algeria 100.
Petroleum	9.9	Nigeria 28, Libya 21, Algeria 18.
Antimony	8.6	S. Africa 96.
Asbestos	8.5	Zimbabwe 51, S. Africa 41, Swaziland 7.
Nickel	7.7	S. Africa 51, Botswana 30, Zimbabwe 19.
Lead ore	5.4	S. Africa 43, Morocco 36, Namibia 13.
Coal	5.2	S. Africa 97.

Zaire–Zambia boundary, accounts for over 80 per cent of Africa's copper production, the remainder also coming from southern Africa. Cobalt is closely associated with copper deposits, mainly in Zaire which is the world's largest producer. And so the treasure house catalogue goes on. Mining is big business in southern Africa and breeds big projects, particularly in transport, which can be, but not always are, used for more diversified economic development.

The coastal states of west and north Africa, Liberia, Sierra Leone and Mauritania have been major producers of iron-ore, but war in Liberia and resource exhaustion in Sierra Leone have caused output to fall away. Railways built from the coast to the iron-ore deposits have not been used for any other purpose and in Sierra Leone were ripped up one the ore was exhausted.

Modern mining is large scale and capital intensive and the mining sector is usually in the hands of multinational corporations, even where some form of nationalization exists. These companies have the capital finance, information, technical know-how, managerial skills and marketing experience to carry out mining efficiently. In dealing with specialist companies the individual state is invariably at a disadvantage. The companies are part of a world trading system which works very much to the advantage of industrial countries. African minerals, exported as raw materials, are turned into manufactured goods in industrial countries. The long-term trend is for raw material prices to fall relative to prices of manufactured goods.

The mining enterprise of an African state is often isolated from the rest of the economy, as if it were an enclave. It is capital intensive, requires very little labour and the massive, fiendishly expensive equipment has to be imported. Observing the open-cast diamond mine at Orapa in Botswana in full production is instructive. The mining workforce is minimal. Each shift requires a bulldozer, a loader and some 40-ton trucks to be driven, but very little else. There are, of course, process workers, surveyors, managers and service workers to keep the plant and machinery operative and to meet the needs of a small mining township isolated on the Kalahari fringe.

On the other hand, the enclave economy concept can mislead because there is a disruptive drain of young men from the traditional economy to the mines. In South Africa, the 'homelands' have been run down as a consequence of the pernicious migrant labour system devised to meet the labour requirements of the mines. Such a dual economy, found in many parts of Africa, comprises two dynamic elements, one feeding off the other.

The rate of mineral extraction is usually controlled by the mining company. The extreme example is in diamonds where the South Africa-based De Beers company, through its Central Selling Organization (CSO), has cornered the world market, including the production from Russia. Many African states accede to the production and marketing dictates of multinational companies because independent action would probably be less profitable.

The top five mainland African states in the GNP per caput league table have economies which are heavily dependent on mineral production. Libya is a one-resource state based on oil. Gabon's modest oil and manganese production, combined with a small population, gives it second place. South Africa's prosperity since 1870 has been based on great and diverse mineral wealth. Algeria is another oil-based economy. The most dramatic example of rags to riches via mineral development is that of Botswana. At independence in 1966 there was virtually no mining and the economy was one of the poorest in Africa in terms of GNP per caput. Then, on the basis of mining for diamonds (from 1971) and for copper and nickel (1974), Botswana began to climb steadily up the league table to seventeenth in 1975, tenth in 1980 and fifth in 1985, a position maintained in 1991.

The other side of the coin is that many of the same African economies are almost totally dependent on minerals. In Algeria, Botswana, Gabon, Guinea, Libya, Mauritania, Nigeria and Zambia, minerals count for more than 90 per cent of exports and most of the countries are dependent on just one mineral. These economies are vulnerable: in the long term to resource exhaustion, in the shorter term to price fluctuation beyond their control.

51 Zambia: a mining economy

Behind marquees on the lawns at State House, Lusaka, in the minutes before independence, agreement was finally reached between Britain and the incoming Zambian government on the transfer of mining rights from the British South African Company to the new Republic of Zambia. A feeling of relief and even euphoria added to the emotions of the occasion. All had come right on the night, the last obstacle to progress was swept away and Zambia's future seemed full of hope. There were potential dangers lurking in the colonial inheritance and, with the hindsight of thirty years, it seems that whatever could have gone wrong did so.

The brave new Zambia was mineral-rich, with the still-expanding copper-belt a hub from which modern development could spread and unite the whole country through enlightened plans for regional development. The new state was heavily dependent on copper and was land-locked, with its colonial transport routes running across the front line between black majority-ruled and white minority-ruled Africa, but even that seemed no major problem as some enthusiastic ex-patriots journeyed in to help via the Benguela railway. The other basic colonial inheritance, of a grotesque, butterfly-shaped boundary encapsulating about five million people of seventy-two different ethno-linguistic groups, did present a challenge but 'one Zambia, one nation' became more than just an empty slogan and provided a vision of the way ahead.

Zambia's land-locked vulnerability was underscored when, following the Rhodesian unilateral declaration of independence (UDI) in November 1965, the net effect of an oil embargo against Rhodesia was to prevent oil from reaching Zambia. Rhodesia interrupted Zambia's traditional access routes to the sea and, although an oil pipeline was built from Dar es Salaam in 1968, the vital bulk copper exports remained at risk. When Western agencies, insistent on financial criteria, would not fund a railway from Dar es Salaam the Chinese did, but low capacity, poor maintenance and congestion at the port limited its effectiveness.

Just as the struggle against dependence on the white south seemed to have been won, Zambia's hopes were dashed by a fall in the world price of copper (set in London) in 1975 and by the closure of the alternative access route, the Benguela railway, by South African-backed UNITA guerillas in Angola. Expenditure exceeded income and international debts began to mount. Drought led to maize shortages, which were made good only after a humiliating approach to South Africa. Peripheral to South Africa's economic core, Zambia tried from 1964 to escape from its orbit but without lasting

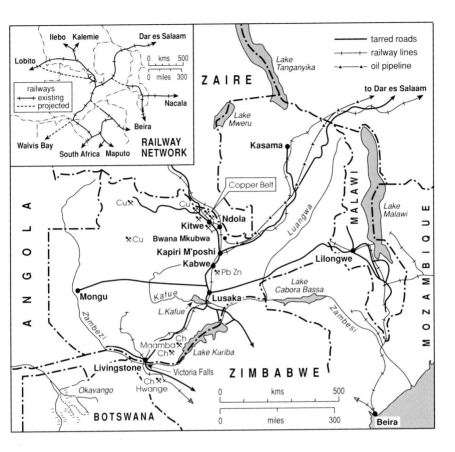

success. The terms of trade deteriorated rapidly and by 1982 Zambia had to export four times as much copper to import the same volume of goods as it did in 1970. Zambia became the first African state to 'reschedule' its international debt repayments. The IMF imposed tough terms, insisting on the ending of food subsidies which provoked popular protest. Zambia broke with the IMF in 1987 but even rising copper prices could not help. The mines had been deprived of investment and production costs had risen as the resource was depleted and mining had to go deeper. A return to the IMF with more realistic restructuring demands prompted the first multi-party elections since independence and in 1991 Kenneth Kaunda was voted out of office. Zambia faced an uncertain future with a fresh but untried government. Once its honeymoon period was over, the new Zambian government faced internal opposition which it felt necessary to meet, in March 1993, with an all-too-familiar state of emergency and detentions without trial of opposition politicians.

52 Manufacturing

Industrial development is often seen as the way for Africa to progress economically. The rationale for this is simple: the industrialized countries of the world are all rich, most non-industrialized countries are poor. Throughout this century the terms of trade have increasingly favoured manufactured goods as opposed to raw materials. This trend is strengthening and the gap between rich and poor, industrialized and non-industrialized is rapidly growing.

In pre-colonial Africa manufacturing industry was small scale and craft-like, such as metal working and pottery. The countries in Africa today which have the largest manufacturing sectors are mainly those which also had large colonial mining sectors, that is, South Africa, Zimbabwe and Zambia. Only in those countries, plus Mauritius and Senegal, does the manufacturing sector exceed 20 per cent of GDP. On the other hand, in fifteen of the thirty-eight African states for which figures are available manufacturing accounts for less than 10 per cent of GDP.

In the most industrialized countries of Africa industrial growth is government policy. Import substitution industries have been encouraged by the erection of tariff barriers. The strategy is not always successful, as such industries often encourage further imports in an insatiable spiral, but it has been widely employed. The motor industry is a good example of such an industry and is found in many different parts of Africa. In Uganda, where the market was too small to support manufacturing, a type of motor industry was set up in a minor but effective way. Imports of motor lorries in a semi-knocked down (SKD) state reduced transport costs. Protected by tariffs and with a minimum of investment in workshop space and basic equipment, important economies were made and local jobs created. The lorries were imported in five sub-assemblies, engine, cab, chassis, drive and wheels. These were bolted together and a locally manufactured body was added. In South Africa, where the local market is very much larger, assembly of completely knocked down (CKD) cars began in the mid-1920s when Ford and General Motors established assembly plants at Port Elizabeth under tariff protection. By 1940 manufacture of 'hang-on' components, namely tyres, batteries and glass, which had a large market in replacements as well as in new equipment, had started near the assembly plants. From the 1960s the South African government, through the Industrial Development Corporation, encouraged the manufacture of motor components, where necessary making direct investment in, for example, an engine block foundry to serve all of the different car plants, so keeping the programme rolling forward.

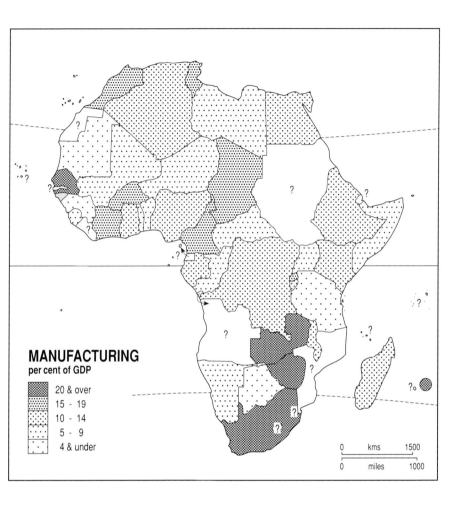

MANUFACTURING
per cent of GDP

- 20 & over
- 15 - 19
- 10 - 14
- 5 - 9
- 4 & under

0   kms   1500
0   miles   1000

Zambia used similar government intervention, albeit on a smaller scale, to encourage industrialization. The Zambian government also adopted the 'growth pole' concept, establishing small industrial sites with factories provided with the necessary service infrastructure to influence the location of industry within the country.

Industrial location is a problem in Africa because of limited infrastructural development. Frequently the main port is the best raw material collection and marketing distribution point. That port, or point of entry in a land-locked state, often has many other functions, such as capital city, which leads to chronic over-congestion, as in Lagos.

53 Wildlife and tourism

A unique feature of Africa is its wildlife. To the world at large the existence and survival of so many animal species is vital. To many African countries wildlife is a major asset, the basis of a thriving tourist industry. But conservation of wildlife is a problem. Conservation must compete against human pressure for land and food; against the ragged armies of African civil wars; against the product of those wars, desperately hungry, displaced people; and against poachers who use the weapons of war with which Africa is saturated to destroy the African elephant for its ivory and the rhinoceros for its horn. Conservation of African wildlife is a world problem but there is a marked reluctance by the richer countries to pay anything significant towards that task. It is left to the individual poor countries of Africa to perform a task which they cannot afford and which in consequence often is not done. Worse than that, pressures from conservation groups in industrialized countries seek to impose solutions on Africa without providing the wherewithal to carry them out and without persuading their own rich governments that the bill should be met.

Most countries in sub-Saharan Africa have set aside large areas as National Parks in which wildlife is preserved and conserved. Some of these National Parks, such as the Kafue in Zambia and the Kruger in South Africa, are larger than states such as the Gambia, Swaziland, Lesotho and Djibouti. The Parks are home to very few people and are infested with such wild animal-borne insects as the tsetse fly which causes sleeping sickness in humans and domestic animals. Within the Parks the wildlife is looked after by teams by wardens who also guard against encroachment by people and the more sinister poachers. They also work to make the Parks available for tourism, which helps to pay for conservation in a neat circularity. Most Parks are well supplied with tourist accommodation ranging from the sumptuous to the primitive. To some degree they are too successful and the large numbers of tourists attracted threaten the well-being of the Parks themselves. Even so, tourism does not cover all of the costs of all of the Parks and in particular it does not meet the high cost of policing vast areas.

The wildlife of Africa is under pressure. The human population explosion in an essentially rural continent increases pressure for land. People encroach on the space of the wildlife. The animals destroy the crops of the people. The people kill the wildlife. It is a deadly sequence of events enacted almost daily all over the continent. Food is often in short supply and wildlife often equals human food. In a trip up country in the Gambia in late 1991 the only wild

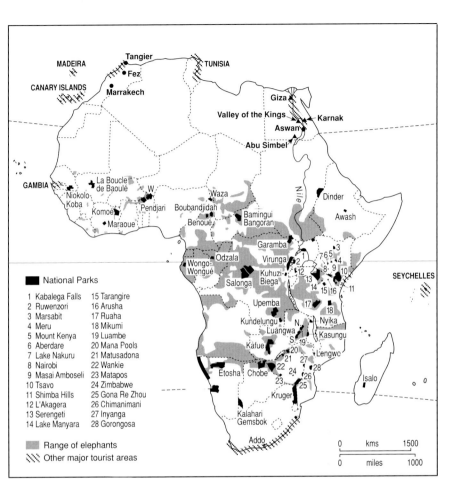

MADEIRA

CANARY ISLANDS

Tangier
• Fez
Marrakech

TUNISIA

GAMBIA
Niokolo-
Koba
Komoé
• Maraoue

La Boucle
de Baoulé
W
Pendjari
Benoué

Waza
Boubandjidah

Giza
Valley of the Kings
Aswan
Abu Simbel

Karnak

Dinder

Awash

Bamingui
Bangoran

Nile

Garamba
Odzala
Wongo-
Wongué
Salonga

Virunga
Kuhuzi-
Biega

1
7 6 5
2
8 9
3
4
10
12
13
14 15 16
11

SEYCHELLES

Upemba

Kundelungu
Kafue
Luangwa
N.
S. 19
20
21
22 Wankie
Etosha Chobe 23
Kruger
Kalahari
Gemsbok
Addo

Ruaha
Mikumi
18
Nyika
Kasungu
L'engwo
27
24
28
26
25

Isalo

o o

National Parks

1 Kabalega Falls
2 Ruwenzori
3 Marsabit
4 Meru
5 Mount Kenya
6 Aberdare
7 Lake Nakuru
8 Nairobi
9 Masai Amboseli
10 Tsavo
11 Shimba Hills
12 L'Akagera
13 Serengeti
14 Lake Manyara

15 Tarangire
16 Arusha
17 Ruaha
18 Mikumi
19 Luambe
20 Mana Pools
21 Matusadona
22 Wankie
23 Matapos
24 Zimbabwe
25 Gona Re Zhou
26 Chimanimani
27 Inyanga
28 Gorongosa

Range of elephants
Other major tourist areas

0   kms   1500
0   miles   1000

animal seen was a 'bambi'-like small antelope which was trussed and bawling its head off on the way to a cooking pot near Basse Santa-Su.

The civil wars of Africa pose another problem. Rampaging armies are often hungry and, with their modern weapons, are able to satisfy that hunger at the expense of wildlife. People displaced by the civil wars sometimes flee into wildlife areas to save themselves. Hungry and desperate they impact adversely upon the wildlife and their habitat.

But of all of the threats posed to African wildlife none is greater than the assault by the poachers. From the 1970s African elephants in particular have been the subject of attention because of the value of their ivory. A market for the raw ivory, which was ready to pay very high prices, existed in the Far East.

In places such as Hong Kong the ivory was carved and prepared for sale all over the world but especially in Japan and the United States. The means of poaching ivory from Africa was also readily available. African civil wars and the willingness of the industrialized nations, including the Soviet Union, to sell vast quantities of arms have resulted in Africa being saturated with the weapons of war. Of particular relevance to ivory poaching is the Soviet automatic rifle, the Kalashnikov AK-47. Whereas the big game hunters of the past were equipped with special heavy 'elephant guns', cumbersome single or double shot rifles, the modern automatic rifle, notably the remarkable AK-47, is a fearsome weapon to be turned on a herd of elephants. With the motivation and the means came the poachers: men, often displaced by war, who were trained in the use of weaponry and hungry for the high rewards available for ivory poaching.

Added to the catalogue of disaster must be the record of some African governments. Some are unable to control poaching because they do not fully control the areas where it is taking place, some because wildlife protection is low on their political agendas and some because they do not have the means available to finance conservation. There are others whose members, from the very top, not only turn a blind eye to what is happening but also lend their support for a share of the spoils.

The toll on African wildlife has been enormous in the three decades of independence. Rhinoceros numbers are estimated to have fallen from about 60,000 in 1960 to under 4000. In 1973 alone, Uganda under Amin lost an estimated half of its 40,000 elephants. In a period when Nairobi was the ivory capital of Africa, Kenya lost half of its elephants between 1970 and 1977. Much Kenyan ivory was illegally smuggled for sale in Somalia where the regime of Siad Barre condoned the traffic. The elephant was also decimated in the Central African Republic, the Congo and Zaire. Illegal exports were carefully disguised by bogus paperwork. In Angola UNITA used ivory to help to pay for their civil war against the MPLA government.

African governments and wildlife conservationists had first to be convinced that the elephant was under grievous threat and then to be encouraged to take appropriate action. Many were reluctant to see a total ban on the ivory trade and so a quota system was introduced. As forecast, it was blatantly abused. In 1989 a total ban on the ivory trade world-wide was introduced effectively to stem the flow of ivory. In the Tsavo National Park in Kenya about 1000 elephants were poached in 1988, in 1990 the number had fallen to a mere fifteen, in 1991 it was apparently nil. In 1992 the ivory trade ban was renewed.

The Convention on International Trade in Endangered Species (CITES) meets every three years. In 1989 a hard fought battle was won, largely against the ivory traders and consumers, to impose the ban on trade in ivory and other elephant parts. In 1992, despite the success of the ban in curbing the African ivory poacher, there was stiff opposition against reimposing the ban from some southern African countries, which argued that they could expand their elephant conservation programmes if they were allowed to fund them from controlled sales of ivory and other elephant products. In countries such as Botswana, Zimbabwe and South Africa elephant poaching has been brought under control to such effect that culling is now deemed necessary. It is claimed that sales of the products of these culls would actually help to improve the position of elephants in Africa. It is a long-running argument which is conducted passionately. The total ban on the ivory trade is construed by some as having been imposed on Africa by outside do-gooders. On the other hand a partial ban, with quotas, has not worked in the past especially in countries where wildlife protection is less well organized.

Although the wildlife of Africa is unique, the continent possesses much more to attract the intercontinental tourist. Sunshine, geographical proximity, cheap accommodation and cheap air travel make attractive holiday packages for literally millions of Europeans each year, mainly in Morocco, Tunisia and the Canary Islands. Extension of this package holiday trade to the Gambia and Senegal is increasing and is becoming a significant earner for those countries.

Elsewhere in Africa tourism depends on unique features to attract visitors to travel further and at higher costs. In Egypt, the Pyramids, Sphinx, Temples and the Nile itself have long brought visitors on the grand scale, as witnessed in the vast hotels of a bygone age in Cairo and Giza. Up-market packages from Europe include the Indian Ocean beaches of Kenya and the Seychelles. But, as with cash crops and raw materials, the tourist trade is not primarily for the benefit of Africans. There are rewards, but they are crumbs from the rich man's table. Air fares and insurance, hotels often managed by non-Africans, food and drink often imported, imported safari vehicles and hotel equipment leave few enough crumbs out of the total cost.

In southern Africa the scenery and the game parks attract some inter-continental visitors. White South Africa generates its own tourists, Swaziland, Lesotho and the 'homelands' have long had a thriving tourist trade, based on casinos and inter-racial sex, for white South Africans seeking respite from the 'Thou Shalt Not' of apartheid and Afrikaner puritanism. In Boputhatswana, Sun City, which is conveniently close to Pretoria and Johannesburg, boasts two five-star hotels which together provide over fifty suites and more than 500 rooms. It is billed as a 'showcase of glitter, gambling and go' which 'rises out of the African bush like an oasis'.

54 Kenya: capitalism and tourism

By virtue of having large tracts of land more than 6000 ft (1829 m) above sea level Kenya was the only equatorial colony to have a large white settler population. The White Highlands north-west of Nairobi were largely given over to white farming and cash crop cultivation. The alienation of this land caused a land shortage among the Kikuyu on the eastern margin of the White Highlands. In the 1950s the Mau Mau troubles, for all the complexity of their origin, were essentially about land. They delayed independence in Kenya until December 1963, after Britain had released Jomo Kenyatta from detention, allegedly for involvement with Mau Mau, to become prime minister and later president.

The abused Kenyatta proved to be a most conservative leader, much praised in the fickle British press that had once vilified him. Under him the Kenyan economy was capitalist free enterprise. Foreign investment was attracted and Nairobi mushroomed, an incongruously high-rise city in the wide open spaces of Africa. In the first two years of independence 14 per cent of the white-held land passed to Africans but most settlers remained in the country. On a broad agricultural base cash crops predominate, led by coffee and tea. The tourist industry is the most highly developed in sub-Saharan Africa, with fine modern hotels in Nairobi and along the palm-fronded Indian Ocean coast. The game parks are so successful in attracting visitors that they threaten their own fragile ecological balance. Investment has also been made in the manufacturing sector mainly at Nairobi and Mombasa, the port of entry. Kenya has few minerals to exploit apart from the soda-ash deposits of Lake Magadi south of Nairobi.

In terms of total GNP Kenya is similar to the Ivory Coast, but a population of almost twice the size halves the per caput measure. Over the decade 1980–90 Kenya experienced the very high average population growth rate of 3.9 per cent per annum and in some years the rate exceeded 4 per cent. The growth rate in GNP over the same period was an average of 4.2 per cent per annum, so that Kenya just kept its head above the water. Kenya's modest economic achievements positively glow alongside those of Tanzania and Uganda, its former partners in the erstwhile East African Community (EAC). Kenya was the linchpin of the EAC, the collapse of which had a great deal to do with the perception of Tanzania in particular that Kenya was profiting greatly from the arrangement at the expense of the other two member states.

The end of the EAC hit Kenya badly. It coincided with the world recession and the collapse of commodity prices, a devastating combination for an

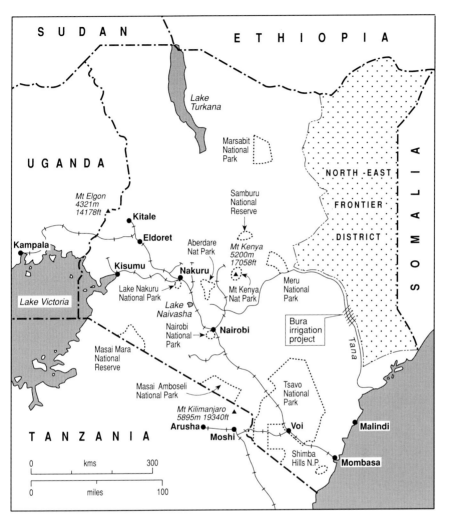

economy dependent on cash crops. The drift of people to the towns continued despite the lack of jobs there. Shanty towns of unemployed have become commonplace and the rocketing crime rate threatens the lucrative tourist trade. The land problem surfaced again as another consequence of rapid population growth. However, the most serious threat to Kenya has been political instability since the death of Kenyatta in 1978. The rule of his successor, Daniel arap Moi, is authoritarian and members of the government are corrupt. Although the regime stumbles from one scandal to another, it has so far resisted pressures to adopt multi-party politics to restore confidence.

55 Migratory labour

In parts of Africa there is disjunction, over distances too great to be covered be mere commuting, between where people live and work. At one level the phenomenon is expressed within, for example, South Africa where many residents of the black homelands migrate to work, usually on short-term contracts in the mines, industries, farms and services of white South Africa. International migration of labour is also found in southern Africa as well as elsewhere in Africa, where people are recruited to work for employers in other countries which have a shortage of labour. For many in North Africa there is intercontinental labour migration to Europe or south-west Asia. At all levels the system is broadly similar. The migrant labourers are mainly able-bodied young men. They travel to their distant jobs, leaving their families behind in the homelands or country of domicile. They support their families by sending them 'remittances' of money from their wages. At the end of their contracts the migrant labourers return home, bringing more of the proceeds of their earnings with them, sometimes in the form of goods not normally obtainable at home. Often they go back to their place of work for further contract periods.

Labour migration usually occurs in lieu of permanent migration by whole families because of restrictions by host countries, including white South Africa, on permanent migration from the black homelands (influx control). The syphoning off of young, able-bodied males from other communities brings to the host countries opportunities in further wealth creation without having to face the full consequences of normal social provision for their workforce. The families of the migrant labourers do not form part of the responsibility for social provision of the host countries and the migrant labourers themselves return home at the end of their contracts. Wages paid to migrant workers are usually low as they generally have low levels of skill and enter the lowest stratum of labour in the host country. The home countries (or homelands) of the migrant workers are usually poor, almost by definition over-populated, and severely under-provided with social infrastructure. Because of restrictions on permanent migration the poor home countries of the migrant workers remain reservoirs of cheap labour whilst the countries that host employment grow comparatively rich partly on the backs of migrant labour.

Perhaps the best-known host country of migrant labour in Africa is South Africa. Soon after large-scale mining started there in 1870 (Kimberley) and 1886 (Witwatersrand), labour supply became a problem. A poll tax was first imposed on Africans in 1895 to drive them into the money economy and the

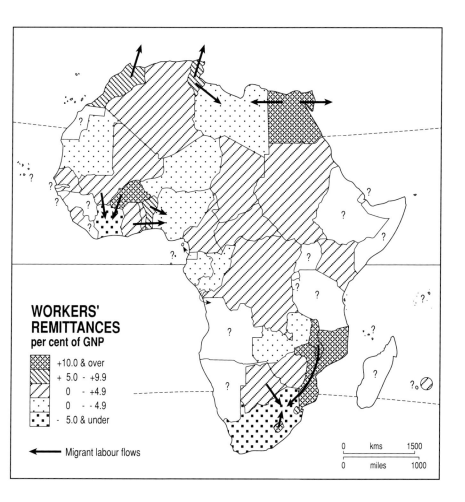

WORKERS'
REMITTANCES
per cent of GNP

+10.0 & over
+ 5.0 - +9.9
0 - +4.9
0 - - 4.9
- 5.0 & under

← Migrant labour flows

0   kms   1500
0   miles   1000

mine-owners set up labour recruiting organizations which spread their net throughout southern Africa. Countries as far afield as Tanzania and Malawi were scoured for labour but the largest numbers were recruited from Mozambique and Lesotho. In west Africa migrant labourers from Niger, Benin, Togo and Ghana are sucked into the large Nigerian economy and from land-locked Burkina Faso to the Ivory Coast and Ghana. Libya imports labour for its oil and associated manufacturing industry from Egypt and Tunisia as well as from outside Africa. Many Tunisian migrants work in southern Europe, notably Italy, and migrants from Algeria and Morocco work in Spain, France and further afield in Europe (refer to Chapter 70, Statistics).

56 Urbanization

In a continent of mainly rural dwellers many of the worst social problems are to be found in African towns. Most African urbanization is recent and problems arise from the over-rapid growth of towns and the favouring of the urban sector at the expense of the rural.

Pre-colonial Africa was also pre-industrial and what towns existed then were essentially pre-industrial. There were the administrative centres of kingdoms and empires from Cairo to Bulawayo. There were trading towns at crossroads of commerce, such as Gao and Timbuctoo on the southern fringe of the Sahara, which commanded routes in virtually all directions. In Yoruba-land the agricultural towns were of a different tradition. South of the Sahara the degree of urbanization was minimal.

Colonialism introduced the administrative capital, usually small in itself but a nucleus for powerful growth. It also brought ports to handle imperial trade, and most doubled as colonial capitals. Settler populations brought urban traditions, founding Salisbury (Harare), Nairobi and Windhoek, while mining areas developed associated towns: Kimberley, Johannesburg, Elisabethville (Lubumbashi) and Ndola. These various colonial implants attracted Africans from rural areas, either drawn in spontaneously by the 'bright lights' or deliberately pulled in by colonial taxes, payable in cash, to perform wage labour. In most colonial towns, influx was carefully controlled. The town itself was planned and laid out in distinctive forms. But colonial urban growth was often considerable: Bulawayo grew from 29,000 in 1936 to 124,000 in 1951, Abidjan from 18,000 in 1933 to 119,000 in 1955.

With independence the move to towns increased. Many colonial curbs were removed, more urban-based jobs were created and the town became the place where it was all happening. In several African countries, Botswana, Swaziland and Tanzania, urban growth rates for the whole period 1960–90 were at a phenomenal level, in excess of 10 per cent per annum. In another twenty-nine African countries the average annual urban growth rate was between 5 and 10 per cent in 1960–90. Low base figure levels and the previous absence of an urban tradition account for these high growth rates but they give a statistical impression of just what is meant by over-rapid urbanization, which is at the root of so many social problems. Much of the increase was in capital cities and the primate city dominates in much of Africa, with the typical rank/distribution curve being 'L' shaped. Conakry, Dakar, Dar es Salaam, Harare, Kampala, Lomé, Luanda, Maputo and Tripoli all contain more than 50 per cent of the urban population of their respective countries.

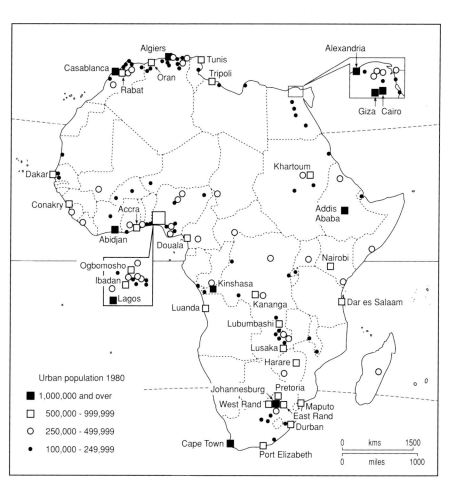

Urban population 1980
- ■ 1,000,000 and over
- □ 500,000 - 999,999
- ○ 250,000 - 499,999
- • 100,000 - 249,999

The attraction of towns outstrips the ability of towns to cope with the inflow of people: basic services such as houses, water, electricity and sewerage cannot be expanded quickly enough. Job creation has not kept pace with the inflow. The response has been not a stemming of the flow but the creation, for want of houses, of shanty towns; and for want of jobs, an informal sector. Urban congestion, chronic in places such as Lagos, slows down economic development. Deprivation and lack of services create serious health hazards and encourage the spread of disease. Already expressing themselves in soaring crime rates, the unemployed, underfed, deprived but somewhat educated urban dwellers could become a major destabilizing influence in a continent that cries out for political stability.

57 Capital cities

African capital cities are usually primate cities, very large relative to other cities, large in absolute terms, very fast growing and faster growing than other cities. All but five African capital cities are the largest urban centres in their respective countries. The exceptions are Porto Novo (Benin) ranked 2, Yaounde (Cameroon) 2, Lilongwe (Malawi) 2, Rabat (Morocco) 2, and Pretoria (South Africa) 4. (Pretoria is the administrative capital of South Africa, with Cape Town the legislative capital and Bloemfontein the judicial capital.)

The African capital city is often the only major urban centre in a state and is frequently many times greater in population than the second-ranking city. The following capital cities are about ten times larger than the next highest ranking town: Luanda (Angola), Bujumbura (Burundi), Bangul (CAR), Djibouti (Djibouti), Conakry (Guinea), Bissau (Guinea–Bissau), Maseru (Lesotho), Bamako (Mali), Maputo (Mozambique), Kigali (Rwanda) and Mogadishu (Somalia). Maputo and Conakry account for over 80 per cent of the urban population of their respective states.

Many capital cities are large in absolute terms. Five have a population of over one million: Algiers, Cairo, Addis Ababa, Abidjan and Kinshasa. Only the least populous states have capital cities of less that 50,000 population: Gaborone (Botswana), Praia (Cape Verde), Moroni (Comoros), Malabo (Equatorial Guinea), São Tomé (St Thomas and Prince Islands), Victoria (Seychelles) and Mbabane (Swaziland).

Capital cities are almost invariably the fastest growing cities in African states. Capitals such as Yaounde (Cameroon), Libreville (Gabon), Abidjan (Ivory Coast), Monrovia (Liberia), Niamey (Niger), Lomé (Togo), Kampala (Uganda), Kinshasa (Zaire) and Lusaka (Zambia), which were all well established at independence, have increased their population by more than five times since 1960. Even in South Africa, fourth-ranked Pretoria is the fastest growing city. In five states the capital is now the largest city but was not so at the time of independence. They are: Gaborone (Botswana), Nouakchott (Mauritania), Khartoum (Sudan), Ouagadougou (Burkina Faso) and Lusaka (Zambia).

In some cases the high growth rate of capital cities is partly explained by changes affecting statistics, by boundary adjustment or by more reliable enumeration. But the major part of growth of capital cities derives from the singular attractiveness of these cities as places of employment, modernity and political power. They often perform a host of functions, being not only capital

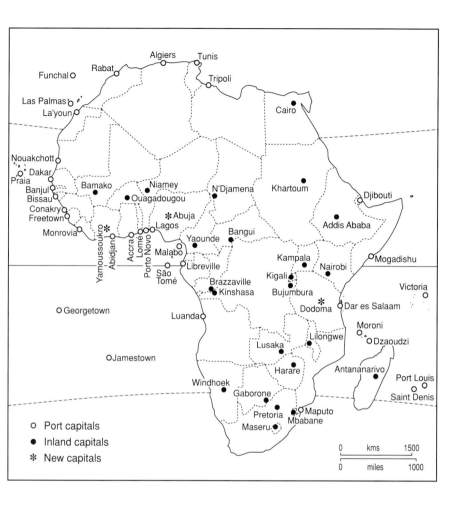

Legend:
○ Port capitals
● Inland capitals
✳ New capitals

city but also port and centre of communications, industry, commerce, education and culture.

Such an overwhelming concentration of functions is, in most cases, part of the colonial inheritance. Colonial capitals, the seat of alien governments, were often located just within the colony, at the point giving best access to the metropolitan country. Hence, along the West African coast, the seaboard states from Senegal to Nigeria, with only two very recent exceptions, have port capital cities. In all, half of the fifty-two African capital cities are ports (fourteen states are land-locked). Only Yaounde, Brazzaville, Cairo, Addis Ababa, Nairobi, Antananarivo, Windhoek, Pretoria, Khartoum and Kinshasa, plus the two new West African capitals of Yamoussoukro and Abuja, are non-seaport capitals of seaboard states. Most have a special reason for being so: they are not colonial creations.

A city that was colonial capital and chief port inevitably accumulated other functions and irresistible attractiveness. The transformation from colonial capital to national capital accelerates growth, as the concentration of functions is reinforced by the addition of the newly acquired instruments and symbols of nationhood and international status. The political function of the capital city in a newly independent state is a virile growth force. A modern seat of government immediately surrounds itself with the seemingly endless personnel of legislative and administrative bureaucracy, political parties and diplomatic representation. Secondary and tertiary growths mushroom, especially with post-independence relaxation of colonially imposed inflex control.

New capitals which were created at independence, such as Nouakchott (Mauritania) and Gaborone (Botswana), have forged ahead to become the largest urban centres virtually from scratch, on the basis of the political function alone. In countries of predominantly rural economic activity, the political/administrative factor is dominant in the process of urbanization. In cities where the political function is one of many, as at Maputo, over-concentration of urban growth leads to chronic congestion and inefficiency.

The marginal location of the colonial capital not only adds the basic function of a port to the capital city but also leaves this concentration of modern sector development remote from the geographical centre of the state. Not many African cities could be more geographically marginal than, for example, Lagos, near one corner of the rough parallelogram of Nigeria, or Lomé, at the corner of a long and narrow, rectangular Togo. Even in some land-locked states the capital city is at or near one frontier, usually at the point most accessible to the former metropolitan power: for example, Bangui (CAR), Maseru (Lesotho) and N'Djamena (Chad). Livingstone, hot and sticky but nearest to Cape Town and London, was the colonial capital of Northern Rhodesia (Zambia) until the higher, cooler, healthier and more central Lusaka was chosen as a new capital in 1931.

The process of economic development which is usually closely associated with the multifunctional capital city is not helped by the marginal location of most African capitals. Spread effects have far to travel, lines of communication are long and tenuous, some regions are incredibly remote. Nor is national unity, a prime concern to so many African states, well served by the geographical remoteness of the capital city.

The imbalance created by the all-attractive, multifunctional, peripherally located capital city has long been recognized, along with the positive, remedial policy option of harnessing the virility of the political capital function in order to stimulate growth and development in more backward, more central regions.

The dynamic capital city force can be relocated in a way in which a port or mineral-based urban complex never can be. But Africa has few Brasilias.

Lilongwe, in land-locked, dependent, conservative Malawi, was the first full post-independence capital city in Africa. The decision to relocate the capital was taken in 1965 but it was not until 1975 that Lilongwe became the capital in succession to Zomba. Lilongwe is 180 miles (290 km) north of Zomba, near the geographical centre and in the widest part of the elongated state. It enjoys good communications by tarred roads and rail and is 85 miles (135 km) from the lakeside. The development of Lilongwe was planned and carried out with substantial financial and technical assistance from a pariah South Africa eager to parade its benevolence. With a population of about 140,000 Lilongwe is successfully contributing to a redistribution of economic development away from the Blantyre region in the extreme south.

Blantyre region in the extreme south.

In 1976 Nigeria decided to move its capital from the port of Lagos over 400 miles (650 km) inland to Abuja, a completely new town in a newly created Federal Capital Territory. Lagos, typically for a colonial capital centred on an island site, is chronically congested. Being located in the middle belt of Nigeria, between the three great regional power blocks, Abuja has a sense of 'neutrality' which Lagos never had and the Nigerians hope that it will become truly the 'symbol of our unity'. The ambitious and costly Abuja project was decided upon during Nigeria's oil-boom years and was too advanced to be abandoned when the downturn in oil came evident.

Affordability is the reason for the delay in fully implementing the decision to move the capital of Tanzania from the port of Dar es Salaam to Dodoma, where the Tanganyika railway crosses the 'Great North Road', 287 rail miles (460 km) west of Dar es Salaam and 267 road miles (427 km) south of Arusha and near the geographical centre of the country. Poverty-stricken Tanzania, unlike Nigeria, cannot throw hundreds of millions of pounds at the scheme and it will take some time to implement.

A more homely variant on the changing capital city theme is that of Yamoussoukro in Ivory Coast. A small provincial town 166 miles (266 km) inland from the Port capital of Abidjan, Yamoussoukro commended itself not just because of its more central location but because it was the birthplace of President Felix Houphouët-Boigny. It is also the site of the president's extravagant cathedral only slightly smaller than Rome's St Peter's.

The inheritance of the colonial capital city has had a harmful effect on many an African state. That few have so far taken positive action to relocate is largely a matter of cost. Advantages apparent in theory are not easily demonstrated in practice. In many states there is no alternative, viable location and for others relocation simply would not be appropriate.

58 Transport

Transport development in Africa has three distinct, if overlapping, phases: pre-colonial, colonial and post-colonial. However, most models of transport development in Africa are basically colonial in concept and so forfeit historical accuracy and current relevance. They also pay little attention to the fact that routes, especially those involving heavy capital investment, are developed with a specific goal in mind. Roads and railways are not built aimlessly across an isotropic surface through an apparently empty interior. Such models lack a logical dynamic and display weaknesses which, enhanced by uncritical reception and dissemination, lead to fundamental misconceptions about African transport development.

Before colonial times in Africa there were rich agricultural areas, concentrations of population, towns and cities and worked mineral deposits. People moved between such places along well-defined routes carrying trade goods over considerable distances. On the southern edge of the Sahara, Timbuctoo and Gao stood at crossroads of trade. Across the Sahara came Mediterranean products and from the South came gold, ivory and slaves. Trans-Saharan and savanna routes long predated European exploration. To reach Timbuctoo Gordon Laing took caravan routes across the desert from Tripoli, while René Caillé obtained passage on a boat sailing down the Niger and returned via the caravan route to Morocco. The route from Bagamoyo to Lake Tanganyika was not hacked out by Burton and Speke but was an old Arab trade route, in a sense colonial as it was developed to exploit ivory and slaves. Kampala–Mengo as a node long predated the arrival of Speke and Grant, and Zimbabwe traded with Sofala before the Portuguese sailed around the Cape. Patently, pre-colonial Africa was not an empty continent.

When the Europeans arrived they established trading posts on the coast and, by offering higher prices for gold, ivory and slaves, contributed greatly to the decline of savanna trade centres such as Timbuctoo. Early European traders rarely ventured far into the interior being content to trade through African middlemen, but when they did it was usually to pre-existing African towns, often for military purposes. In present-day Ghana the British, under Sir Garnet Wolseley, moved from the coast to attack Kumasi, the capital of the troublesome Ashanti kingdom. As the expeditionary force advanced, Wolseley's troops built a military road from Accra to Kumasi, a route later followed by the railway that was built under the British colonial administration in 1923. In Nigeria British penetration was also to predetermined

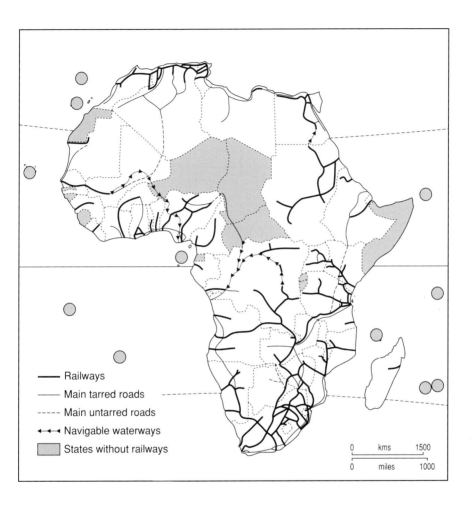

Railways
Main tarred roads
Main untarred roads
Navigable waterways
States without railways

| 0 | kms | 1500 |
| 0 | miles | 1000 |

points, to pre-existing African cities such as Ibadan and Benin and later to Kano and Sokoto. Colonial transport routes were built from colonial ports to pre-colonial places and were not developed in a virgin wilderness.

In southern Africa the colonial phase may be usefully subdivided into pre-industrial and industrial, the latter dating from 1870 and the discovery of diamonds at Kimberley. Pre-industrial colonial transport was mainly ox-wagon, a leisurely but effective means of transport from the ports to the few inland centres. Accounts of such travel survive in sources as well known as

Livingstone's *Missionary Travels* (1857). The interior had little to attract investment in any means of transport other than ox-wagons which were adequate for carrying the wool, hides and ivory produced in the interior. Penetration by railway was discouraged by the outward-facing Great Escarpment and, in the south-west, by the ranges of the Cape Fold Mountains also parallel to the coast. By 1870 there were just 69 miles (110 km) of railways in South Africa, mainly linking Cape Town with Stellenbosch, Paarl and Wellington across the Cape Flats.

The discovery between September 1870 and June 1871 of four large diamond 'pipes' within a radius of two miles (3 km) at the heart of the subcontinent transformed transport development in southern Africa. The Cape Government took over the existing private railways and, under political pressure from the regions, simultaneously built lines towards Kimberley from the three ports of Cape Town, Port Elizabeth and East London. A new 'Cape gauge' of 3 ft. 6 in. (1.06 m) was adopted to ease the engineering and financial problems of breaching the mountains but it was not until November 1885 that the railway reached Kimberley simultaneously from Cape Town and Port Elizabeth via De Aar junction.

In 1886 the Witwatersrand gold fields were discovered, also localized and deeper in the interior. Although a railway had been built to Ladysmith from Durban, the first line to reach Johannesburg was an extension of the Cape line via Bloemfontein. Construction over the relatively flat high veld was quicker and cheaper than climbing the escarpment afresh, but the Natal line was held at the Transvaal border for political reasons to allow completion of the Transvaal's own, non-British line from Lourenço Marques (Maputo). The Cape–Bloemfontein line had been allowed to enter the Transvaal in 1893 only after considerable political and financial pressure had been exerted on President Kruger by Cecil John Rhodes, then Prime Minister of the Cape Colony.

From 1890 Rhodesia (Zimbabwe) became the next node in the continental interior of southern Africa. Although the short, direct Beira route was the looked-for means of access, the Matabele campaign of 1896 and the rinderpest pandemic, which disrupted ox-wagon transport at a critical juncture, led to the urgent mile-a-day extension of the spinal railway from Mafeking (Mafikeng) to Bulwayo, which was completed in November 1897. The Beira line, slowly struggling through fevered swamp and over rugged escarpment, was completed to Salisbury (Harare) in 1900.

Further north the newly discovered mineral wealth of Broken Hill (Kabwe) and Katanga (Shaba) became the next nodes of attraction, reached by the spinal railway in 1905 and 1920 respectively. The all-Belgian line from Port Francqui (Ilebo) on the Congo river system to Elisabethville (Lubumbashi) was completed in 1926, in advance of the Benguela railway from Lobito, also to Kantanga, in 1928.

The South African campaign against the Germans in South West Africa (Namibia) in 1915 led to the Cape network being connected from De Aar to the German rail system, which was upgraded to Cape gauge, and hence to the two ports of Lüderitz and Walvis Bay.

Southern Africa has by far the largest single rail network in Africa, over 20,000 route miles (32,000 km) in twelve different countries, all at the Cape gauge. The densest part of the network is in the richest country, South Africa, which has over 13,500 route miles (21,560 km). The rail network does not merely serve important mineral-based nodes but also encourages development between nodes: hence the concentration of economic development along the Durban–Johannesburg corridor, the Harare–Bulawayo axis and the Zambian 'line of rail'. Originally a colonial creation designed to facilitate mineral exploitation and to strengthen political domination, the rail network has become the infrastructural framework for the distribution of economic development. In the 1980s it provided white South Africa with the means of exerting economic and political hegemony over neighbouring states, despite important post-independence railways which were built specially to help to break that dominance.

Post-colonial or independent transport development began in 1894 when the Transvaal's direct link with Lourenço Marques was built in an attempt to escape the British stranglehold on the Boer Republic. In 1955 Rhodesia opened a direct rail route to Lourenço Marques via Malvernia (Chicuala-cuala) in order to maintain economic independence from South Africa. However in 1974, when Mozambique was about to become independent and white Rhodesian links with the sea were threatened, Rhodesia quickly completed its direct rail link with South Africa via Beit bridge, having delayed doing so for almost fifty years. The most spectacular post-colonial railway in Africa is the 1050 mile (1680 km), Cape gauge Tanzania–Zambia (TAZARA) railway from Zambia to Dar es Salaam. It was planned specifically to help Zambia to escape the economic and political clutches of white Rhodesia and South Africa. Less ambitious but also built to give a land-locked state alternative access to the sea is the rail link between Malawi and the northern Mozambique port of Nacala. All of these lines reflect the fact that the decisions to build them were taken in the independent capital cities of African (and Afrikaner) land-locked states. They are not lines likely to have been built by the colonial powers.

In East Africa three colonial lines from the ports of Mombasa, Tanga and Dar es Salaam were joined to form a single network at a metre gauge. The original lines were built for strategic colonial reasons to Lake Victoria and

Uganda, to the rich agricultural area near Mount Kilimanjaro and to Lakes Tanganyika and Victoria respectively. The first two were joined in 1916 by South African forces under General Smuts, fresh from linking the Cape– South West African systems. The final link deliberately to make a single network significantly came after independence in 1964. The port of Dar es Salaam serves both east and southern African rail networks which are of different gauge.

Elsewhere in Africa, except in Maghreb where a lateral line runs from Marrakech to Tunis, the railways are essentially colonial: short lines from the coast inland to mines, larger towns or rich agricultural areas with no lateral linkages. Three of them, from Dakar, Abidjan and Djibouti, do cross international boundaries. In Liberia and Guinea individual lines are of different gauge, making any lateral linkage difficult. Post-colonial additions have also extended some railways into remote areas to assist regional development, for example, the lines to Maiduguri in north-eastern Nigeria, to Packwach in Uganda and the Trans-Gabon railway. Ten territories in continental Africa have no railways at all.

Although railways as the means of trunk transport are well suited to Africa, with its great distances and extensive plateau surfaces, road building has gained momentum in the post-colonial period. Many trunk roads merely duplicate routes already served by rail, often from ports into the interior, and others connect places previously served by river transport, for example, from Banjul along the south bank of the river Gambia to Basse Santa-Su, the first 125 miles (200 km) of which to Mansa Konko is tarred using a matrix of sea shells rather than stone chippings. But increasingly tarred roads are being used to reach those parts previously not connected by modern transport to the capital city, for example, the Trans-Gambia Highway from Dakar across the Gambia to the previously remote Casamance region of Senegal, from Lusaka via Mumbwa to the upper Zambesi valley and from Nairobi northwards towards the Ethiopian border.

The pattern of international transport in Africa is still basically colonial, reflecting the continuing pattern of African trade which remains mainly with industrialized countries rather than being intra-African. The painfully slow progress of the long-mooted trans-African highway project illustrates the lack of urgency felt in Africa for such links. The Trans-African Highway Authority was formally inaugurated in 1981, ten years after it was proposed by the ECA. More than a decade later, the project is nowhere near completion.

If Africa is ever to free itself from a world trading system in which it occupies a subservient and dependent role it must begin to develop an international, continental infrastructure which is capable of materially assisting alternative trade patterns. Intra-African trade can be expanded only when the means of pursuing that trade are in place. Once in place it is equally important that they are maintained, road surfaces repaired, locomotives serviced, and that the railways in general run efficiently.

59 Energy resources and utilization

Africa is well endowed with energy resources and vast areas not yet prospected in detail will yield much more. Africa produces more than twice the amount of energy that it consumes and has by far the lowest per caput consumption of energy of any continent. Yet several African states bear a crippling cost of energy imports. The distribution of energy resources is far from even and the few states with a large surplus of energy production over consumption sell their produce on world markets at prices which their poor neighbours in Africa without energy resources cannot afford.

In 1989 Africa produced 646 million tonnes of coal equivalent and consumed 255 million tonnes. Yet only ten African states produced more energy than they consumed: forty-two were net importers of energy. During the 1980s African consumption of energy rose more quickly than African production of energy (by 37 per cent compared with 4 per cent). The imbalance of energy production and consumption was such that Algeria, Angola, Congo, Gabon, Libya and Nigeria all produced more than five times the amount of energy that they consumed. Africa accounted for almost 10 per cent of world oil output but 89 per cent of that came from Nigeria, Libya, Algeria, Egypt and Angola. South Africa produced 97 per cent of all African coal: 180 million tonnes in 1990. To make up for its lack of oil and to give some leeway against an official, though ineffective, oil embargo South Africa converts low-grade coal into oil at the Sasol plants. Low pithead prices enable South African coal to compete on world markets despite high transport costs and in 1989 over 40 million tonnes were exported. Zimbabwe, Zambia and Morocco are other African coal producers but their scale of production is very small and mainly for domestic production. Hydroelectricity in Africa comes mainly from large dams such as Akosombo, Aswan, Cabora Bassa, Inga, Kafue, Kariba and Owen Falls. The output from several of them is in excess of local consumption and electricity is sold to neighbouring states. Virtually all of the uranium produced in Africa is exported and only South Africa has a nuclear power station.

Oil often distinguishes rich from poor in Africa. The very few African states with a favourable balance of payments include oil producers, and the trade deficit of many of the other states is a result of the high cost of oil imports. The poverty of many African states shows up as low energy consumption but even lower production. In the poorest countries of the Sahel firewood as a source of energy becomes a significant element in the delicate environmental balance.

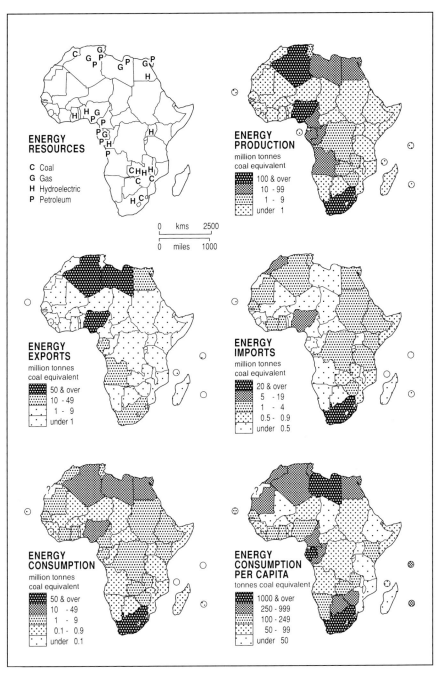

ENERGY
RESOURCES

C Coal
G Gas
H Hydroelectric
P Petroleum

0   kms   2500
0   miles   1000

ENERGY
PRODUCTION

million tonnes
coal equivalent

100 & over
10 - 99
1 - 9
under 1

ENERGY
EXPORTS

million tonnes
coal equivalent

50 & over
10 - 49
1 - 9
under 1

ENERGY
IMPORTS

million tonnes
coal equivalent

20 & over
5 - 19
1 - 4
0.5 - 0.9
under 0.5

ENERGY
CONSUMPTION

million tonnes
coal equivalent

50 & over
10 - 49
1 - 9
0.1 - 0.9
under 0.1

ENERGY
CONSUMPTION
PER CAPITA

tonnes coal equivalent

1000 & over
250 - 999
100 - 249
50 - 99
under 50

60 Harnessing Africa's rivers

Water is the critical resource of Africa. Large areas of the continent are deficient in water to the extent that cultivation is impossible. Other areas have a marked excess of water, often seasonal, leading to waste of a scarce resource and sometimes to destruction of life and land: drought and flood are characteristic of much of Africa. Yet irrigation and hydroelectric potential is very high. Therefore there is a premium on controlling and harnessing Africa's rivers.

Irrigation is most effective in those areas of Africa lying between the high rainfall equatorial zone and the deserts, an 'irrigation belt' from Senegal, through the Sahel, the Sudan, the Horn and east Africa, and most of southern Africa. Over this vast area low rainfall and high evapotranspiration conspire so that cultivation is made possible or at least is greatly improved by irrigation. Availability of water is critical: some small schemes use pumped ground water but the overwhelming majority of irrigation projects use water from rivers and lakes. The largest schemes are to be found where great rivers such as the Nile, Niger and Orange flow through the irrigation belt.

The high plateau surfaces of Africa, often saucer-like in structure, present an escarpment to the sea. Over and through this escarpment the rivers of Africa plunge, their volume and head of fall providing the continent with the world's greatest potential in hydroelectric power, much of it in the same great river valleys that also have high irrigation potential.

Most African rivers have a highly seasonal regime. River courses are often completely dry in one season, full of raging torrents in another. River mouths are sand-dune blocked lagoons for most of the year but then, in flash floods, spew out vast quantities of liquified topsoil to discolour the sea for miles from the shore. River flow is erratic, river flood is devastating. Protection can only come from careful control, an integrated system which involves not only the massive concrete structures of Aswan and Kariba but also small earth dams on minor streams and headwaters. Some idea of the scale of river control that is needed in Africa may be gained from the fact that in South Africa alone there are over half a million dams of all sizes. The process of controlling Africa's rivers has a long way to go.

Many large dams have been built in post-independence Africa but few since the 1970s. Most were primarily to produce cheap electricity, but some were also for irrigation and other purposes. Dams are beloved of politicians, national plan-makers, financiers and aid-donors alike. They are potent symbols of economic virility and political prestige; they are clearly visible,

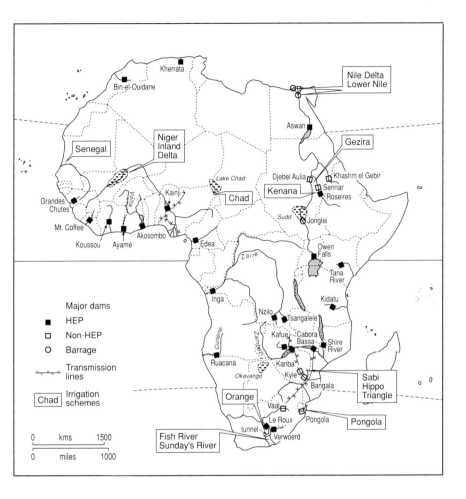

Major dams
■ HEP
□ Non-HEP
○ Barrage
⊦⊦⊦⊦ Transmission lines
| Chad | Irrigation schemes

0 kms 1500
0 miles 1000

concrete and finite projects, demonstrably a basis for future economic and social development.

Not surprisingly, decisions to build dams often have been highly political and controversial. The Kariba dam was a symbol of federation between the Rhodesias, because it straddled the boundary between them, and so the technically superior Kafue site, near the copperbelt consumers, was passed over. Cabora Bassa, built in Portuguese Mozambique, was a manifestation of South Africa's outward-looking policy. Completion of the project coincided with the independence of Mozambique, but electricity flowed along the long transmission lines to Pretoria despite that. Kwame Nkrumah of Ghana was overthrown just before his great Akosombo dam was completed in 1966. The then high cost of the dam helped to cause his fall. Western reluctance to finance the Aswan high dam in Egypt opened the door to Soviet influence in Africa and precipitated the Suez crisis of 1956. Although many African dams have been one-off projects, some do form the basis of international co-operation. Akosombo electricity is supplied to Togo and Benin as well as Ghana, and Kariba supplies both Zambia and Zimbabwe. Yet the only river basins to have been harnessed on anything approaching a fully co-ordinated system are the Nile and the Orange.

For thousands of years people have understood the character of the Nile and have used it to create a great civilization based on intensive cultivation in an area surrounded by desert. Today, from the Owen Falls dam near Lake Victoria to the delta, people have sought to control and harness the Nile through modern technology. The first modern dam on the Nile was completed at Aswan in 1902 to store water for additional irrigation in the lower Nile valley and delta. It was designed to help to control the Nile's flood, which saw the September discharge of the river at about ten times the volume of the April discharge.

The flood comes mainly from the Ethiopian highlands via the Blue Nile which has an annual discharge about twice that of the White Nile. In 1925 the Sennar dam on the Blue Nile was completed to control that flood and to start the Gezira irrigation scheme, the largest in Africa. The first water agreement between Egypt and the Sudan in 1929 gave 5 per cent of the Nile's water to the Sudan. A new post-independence agreement in 1959 adjusted the Sudan's share to 20 per cent following a major expansion of the Gezira scheme, the Managil extension, which opened in 1958. Meanwhile a further measure of flood control and water conservation was achieved by the construction of the Djebel Aulia dam on the White Nile, 30 miles (48 km) above Khartoum. This dam was to pond back the more regular flow of the White Nile when the Blue Nile was in flood, so flattening out the flood peak to give Egypt the opportunity to conserve more water at Aswan.

Such international co-operation was largely a benefit of almost all of the Nile valley being under British influence, if not direct colonial rule. It enabled a detailed plan for water conservation and control, irrigation and hydroelectric power to be drawn up for the whole of the Nile basin and laid the basis for the co-operation and co-ordination which has characterized water development in the Nile valley in recent years.

In 1959 the Owen Falls hydroelectric dam near the Nile's exit from Lake Victoria was completed in order to provide cheap electricity for economic development in Uganda and also Kenya. The Aswan High Dam, which was completed in the late 1960s 4 miles (6 km) above the first Aswan dam, was both controversial and symbolic. Doubts were expressed at the wisdom of creating

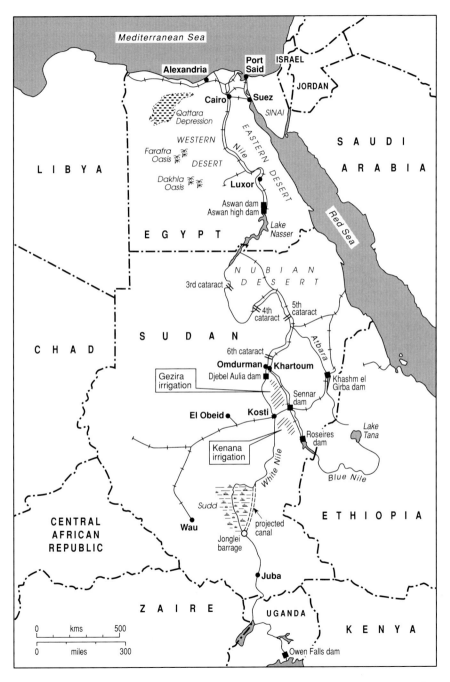

Mediterranean Sea

Alexandria

Port Said

ISRAEL

JORDAN

Cairo

Suez

SINAI

Qattara Depression

WESTERN

Farafra Oasis

DESERT

Dakhla Oasis

Nile

EASTERN DESERT

Luxor

Aswan dam
Aswan high dam

Lake Nasser

L I B Y A

E G Y P T

S A U D I

A R A B I A

Red Sea

N U B I A N   D E S E R T

3rd cataract

4th cataract

5th cataract

C H A D

S U D A N

6th cataract

Omdurman **Khartoum**

Atbara

Gezira irrigation

Djebel Aulia dam

Khashm el Girba dam

Sennar dam

El Obeid **Kosti**

Lake Tana

Roseires dam

Kenana irrigation

White Nile

Blue Nile

CENTRAL AFRICAN REPUBLIC

Sudd

Wau

projected canal

Jonglei barrage

E T H I O P I A

Z A I R E

Juba

UGANDA

K E N Y A

Owen Falls dam

| 0 | kms | 500 |

| 0 | miles | 300 |

the world's second largest man-made lake in one of the hottest, sunniest places on earth, at a site where the shallowness of the lake would add to the ratio of evaporation to storage capacity. Other experts confidently forecast that the dam would silt up very rapidly. But the real controversy, and perhaps the real source of at least some of the technical doubts, concerned its cost, over £400 million, and the source of the funding. The West turned down the opportunity to finance and build the project, anticipating Nasser's fall rather than that the Soviet Union would step in. Not for the last time in modern Africa was the West to misunderstand the motivation and determination of African leaders, and to fail to appreciate the political significance of a project which, assessed in narrow economic terms, might not appear to be a viable proposition. The High Dam became as much a symbol of Egypt's revolution and independence as Egypt's control of the Suez Canal itself. Aswan became the gateway for the Soviet Union's entry into African affairs, though it remained their only venture of that kind in Africa. The Aswan High Dam was

Selected major dams in Africa

Country	Dam	River (basin)	MW	Date
Angola	Cambambe	Cuanza	260	–
Cameroon	Edea	Sanaga	270	1966
Egypt	Aswan High	Nile	2100	1970
Ghana	Akosombo	Volta	792	1966
Ivory Coast	Kossou	Bandama	180	1973
Mozambique	Cabora Bassa	Zambesi	2000	1974
Nigeria	Kainji	Niger	960	1968
South Africa	Vaal	Vaal (Orange)	–	1928
	H.F. Verwoerd	Orange	320	1971
	P.K. Le Roux	Orange	220	1977
Sudan	Sennar	Blue Nile	–	1925
	Djebel Aulia	White Nile	–	–
	Roseires	Blue Nile	–	1968
	Khashm el Girba	Atbara (Nile)	–	–
Tanzania	Kidatu	Nkulu (Rufiji)	100	1975
Uganda	Owen Falls	Victoria Nile	150	1954
Zaire	Inga I	Zaire	350	1972
	Inga II	Zaire	1000	1977
Zambia	Kafue	Kafue (Zambesi)	750	1972
	Iteshiteshi	Kafue (Zambesi)	–	1976
Zambia/Zimbabwe	Kariba	Zambesi	1600	1960

Note: The installed hydroelectrical capacity (MW) shown is the latest known capacity, which may not be the original installed capacity.

completed ahead of schedule to become the linchpin of the Egyptian economy. It has permitted a 20 per cent more intensive use of the previously irrigated land and has allowed 5–10 per cent more land to be brought under irrigation. It provides more than half of Egypt's electricity and protects the country from flood. Evaporation loss from Lake Nasser is high, about 10 per cent of annual flow, but silting is less than anticipated. Salinity has not been a major problem in the irrigated lands but imports of fertilizer have increased. Erosion of the delta coast is experienced with the attendant dangers of sea-water pollution but, on balance, the project has been an outstanding success.

An interesting offshoot of the High Dam project, which underlines the co-operation between Egypt and the Sudan, is the new dam and irrigation scheme at Khashm el Girba on the Atbara tributary of the Nile in the eastern Sudan. Here, people displaced by the creation of Lake Nasser from the Sudanese area around Wadi Halfa have been resettled around New Halfa at a cost, borne by Egypt, of £15 million. At Roseires a new dam was built across the Blue Nile in 1968 to provide additional water storage, flood control and a hydroelectric capacity. It also enabled the creation of the Kenana irrigation scheme of about 1 million acres (400,000 ha). Originally planned for commercial cotton production, the irrigation schemes are the economic heartland of the Sudan and are now also producing groundnuts and wheat as well as subsistence crops for the farmers themselves. In addition to the dams and barrages the Nile is also harnessed to many pump irrigation schemes, the possibilities for which have increased where water levels have been raised by the dams. Plans to increase the flow of the White Nile through the swamps of the Sudd have been set back by the continuing civil war in the southern Sudan. Much remains to be done before even the Nile is fully harnessed, but the need for river basin planning is amply demonstrated and the benefits from international co-operation are plain to see.

The only other African river basin whose potential has been realized to a similar degree is the Orange in South Africa. Its major tributary, the Vaal, flowing westwards from the eastern plateau edge, has long been carefully conserved to provide water for Johannesburg and the other Reef towns and, lower down, for the Vaal–Harts irrigation scheme. By 1974 the long-feared water supply constraint on the growth of the Witwatersrand approached as demand for Vaal water equalled reliable yield, but in the same year a project designed to augment the water yield of the Vaal came into operation. East of the Drakensberg escarpment water is in plentiful supply in the Tugela river basin, so the Tugela has been linked to the Vaal by a pumping scheme which raises Tugela water through a 1660 feet (506 m) vertical lift to the Vaal basin. There the water is stored in the new Sterkfontein earth dam. The project is designed to increase the net yield of water from the Vaal by about 25 per cent.

Future plans to enhance further the water supply in the Vaal basin are contained in the Lesotho Highlands water project, which was agreed between the governments of South Africa and Lesotho in late 1986. This allows for water to be dammed near the sources of the Orange river in Lesotho and then taken by tunnel to the headwaters of the Vaal.

The Orange River Project also transfers large quantities of water from one river basin to another. This project extends irrigation in the lower Orange river valley, gives greater control over a river where peak flow is sixteen times minimum flow, provides hydroelectricity and transfers water from the Orange to both the Fish and Sunday's rivers below the south-eastern escarpment for

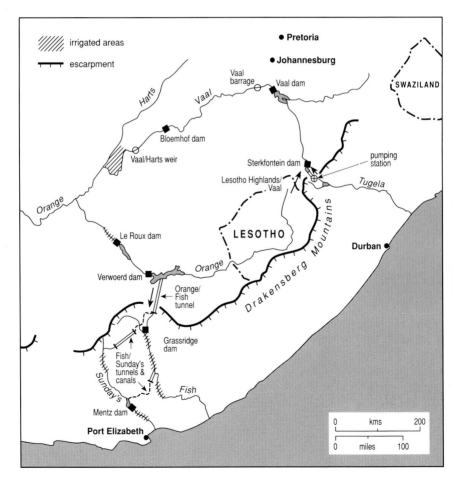

the purposes of irrigation and of supplying the Port Elizabeth/Uitenhage urban-industrial complex. Although the main dams and tunnels were completed by the late 1970s, work continues on irrigation canals and smaller tunnels.

61 Regional economic groupings

Having failed to take up Kwame Nkrumah's challenge to unite, African states have spent the time since independence in setting up regional economic groupings in attempts to overcome both the colonial inheritance of political balkanization and the effect of an inferior place in the world economic system. Small political units make weak, dependent national economies which are easy prey to a neocolonialism which promotes an economic system in which Africa is near the bottom of the pile. Many different regional groups have been formed to face the problems of diseconomies of scale and to promote intra-regional trade. There have been failures but some have survived though progress has been very slow.

At independence the most promising grouping was in east Africa. After the First World War, the British took advantage of the fact that they administered Kenya, Tanganyika, Uganda and Zanzibar and so were able to create many common services for the four territories, ranging from railways and harbours, customs and excise, posts and telegraphs through to East African Airways. Building on this 'very wide basket of joint activities', the independent countries of east Africa came together in 1967 to form the East African Community. It lasted a mere ten years because what many regarded as the true strength of the Community, the joint services, were not in the end able to balance out advantages to the different partners. In particular Kenya seemed to benefit in such areas as incoming industrial investment and no adequate means was found of compensating Tanzania and Uganda. There was ideological friction between Kenya and Tanzania, and the accession of Amin in Uganda provided a different political *coup de grâce*.

French colonial groupings tended to survive independence because it came simultaneously to territories, many of which needed economic support. *L'union Douanière et Economique de L'Afrique Centrale* (UDEAC) succeeded AEF as a customs union in 1966 and also aimed to promote the economic development of member states. It is less ambitious and its services less centralized than first conceived but it has survived. *Organization Commune Africaine et Mauricienne* (OCAM) has had many defections and now concentrates on technical and cultural co-operation. *Communauté Economique de L'Afrique de l'Ouest* (CEAO), the successor to AOF, aims to lower customs duties between members and to promote economic development, objectives shared by the Economic Community of West African States (ECOWAS) to which all of the CEAO states belong.

ECOWAS, formed in 1975, comprises fifteen west African states from

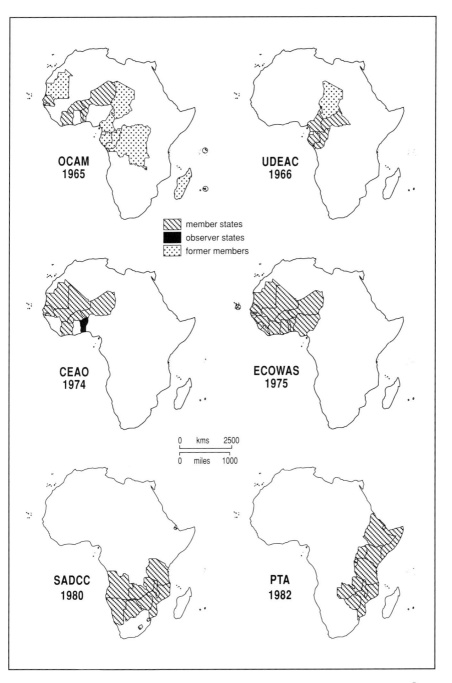

OCAM
1965

UDEAC
1966

member states
observer states
former members

CEAO
1974

ECOWAS
1975

0   kms   2500
0   miles   1000

SADCC
1980

PTA
1982

British and French colonial backgrounds plus Liberia. Its aims are to move trade towards customs union and to promote regional co-operation through commissions for agriculture, industry, energy, trade, transport and telecommunications. Progress has been chequered and closely related to the prosperity of Nigeria, the dominant economic and political power within ECOWAS. The political dimension of ECOWAS came to the forefront in August 1990 when a multinational military force was sent in its name to intervene in the desperate situation in Liberia. The main initiator was Nigeria, whose contribution to the expedition was the largest, but many ECOWAS members contributed, even the tiny Gambia whose President Jawara was strongly in favour of intervention particularly during the period of his Chairmanship of ECOWAS. The move has not brought peace to Liberia and to some extent has further muddied already murky waters, which is not surprising as not all ECOWAS members united behind the purposes of the expedition. But it is significant that ECOWAS acted positively as a political organization, fulfilling a role perhaps more appropriate to the OAU. Success in the Liberian operation is a long time coming but Nigeria has persevered. Recent moves by the various rebel forces, which have resulted in refugees flooding into not only Sierra Leone but also the Ivory Coast, could have the effect of uniting ECOWAS members and could lead to the success of the whole operation. In that case ECOWAS would be transformed and Nigeria's standing would be greatly enhanced. Ironically, this could be a step too far for other member states. In addition to the clear economic dominance of Nigeria, its importance might become so great as to cause other member states to shy away and so endanger the very survival of ECOWAS.

The Southern Africa Development Co-ordination Conference (SADCC) was set up in 1980 by the front-line states to encourage economic development and co-operation independent of a hostile South Africa. Much of the work of the organization was negated by South Africa's destabilization of its neighbours. Roads, railways, bridges and harbour installations were sabotaged by the white South Africans and their proxies, the MNR in Mozambique and UNITA in Angola. During SADCC's first decade, transport was the main focus of its efforts and the organization served well as an aid umbrella, being particularly successful in attracting aid from the Nordic countries. The changes taking place in South Africa led SADCC leaders to state in 1992 that they would welcome participation in SADCC of a majority-ruled South Africa. But until now one of the main thrusts of SADCC has been development which is *independent* of South Africa. The economic strength of a friendly South Africa might be made more difficult for its neighbours to withstand than the brutal hostility of the apartheid state.

Parallel with SADCC is the Preferential Trade Area (PTA) which came into force in 1982. It embraces a wider area in southern and eastern Africa than SADCC, from Djibouti to Lesotho. It aims to promote intra-regional trade and joint action by member states for the production of certain goods, for services, including financial, and for resource development, in an attempt to integrate the national economies into a regional economic community. Intra-regional trade is very low compared with the amount of extra-regional trade because exports are mainly primary products and imports manufactured goods. In a difficult first decade the PTA has made little progress but some structures which have been put in place could lead to useful developments, particularly when membership is extended to South Africa. The price paid could be dominance by one power at the expense of the others, but at least it would be an African power.

62 Further reading

Binns, J.A. (1992) 'Traditional agriculture, pastoralism and fishing', in M.B. Gleave (ed.) *Tropical African Development*, London: Longman, 153–91.

CIMADE, INODEP, MINK (1986), *Africa's Refugee Crisis*, London: Zed.

Douglas-Hamilton, I. and O. (1992) *Battle for the Elephants*, London: Doubleday.

George, S. (1988) *A Fate Worse Than Debt*, London: Penguin.

Gill, P. (1986) *A Year in the Death of Africa*, London: Paladin.

Graham, R. (1982) *The Aluminium Industry and the Third World*, London: Zed.

Griffiths, I. Ll. (1986) 'Web of steel: the development of railways in southern Africa', *Geographical Magazine* 58 (10): 512–17.

Griffiths, I.Ll. (1992) 'Mining and manufacturing in tropical Africa', in M.B. Gleave (ed.) *Tropical African Development*, London: Longman, 223–49.

Griffiths I. Ll. and Binns, J.A. (1988) 'Hunger, help and hypocrisy: crisis and response to crisis in Africa', *Geography* 73 (1): 48–54.

Harrison, P. and Palmer, R. (1986) *News out of Africa: Biafra to Band Aid*, London: Hilary Shipman.

Hodder, B.W. and Gleave M.B. (1992) 'Transport, trade and development in tropical Africa', in M.B. Gleave (ed.) *Tropical African Development*, London: Longman, 250–83.

Kibreab, G. (1985) *African Refugees: Reflections on the African Refugee Problem*, Trenton, NJ: Africa World Press.

Lanning, G. (1979) *Africa Undermined*, London: Penguin.

O'Connor, A. (1983) *The African City*, London: Hutchinson.

Onimode, B. (1989) *The IMF, the World Bank and the African Debt: The Economic Impact*. London: Zed.

Potts, D. (1985) 'Capital relocation in Africa: the case of Lilongwe in Malawi', *Geographical Journal* 151 (2): 182–96

Riddell, R.C. (1990) *Manufacturing Africa: Performance and Prospects of Seven Countries in Sub-Saharan Africa*, London: Currey.

Timberlake, L. (1985) *Africa in Crisis: the Causes, the Cures of Environmental Bankruptcy*, London: Earthscan.

E The South

63 South Africa: apartheid and its demise

In 1948 the white-only electorate of South Africa voted into power the (Afrikaner) National Party, somewhat to the surprise even of its then leader Dr D.F. Malan. It has remained in government ever since through a succession of six different leaders. A narrow, right-wing party, it stood for *Baaskap*, keeping blacks in their place. During its first ten years in office the government used every device systematically to eliminate effective opposition and, through the work of several commissions, did much spade work towards the full flowering of *apartheid*, Afrikanerdom's blueprint for survival, that was to follow. The chief architect of apartheid was Dr H.F. Verwoerd who succeeded to the premiership in 1958. Over eight years, until his assassination in 1966, he put South Africa on a new course. In 1961, after a whites-only referendum, Verwoerd led South Africa to a republic outside the British Commonwealth and so ended the Anglo–Boer war of 1899–1902. The Promotion of the Bantu Self-Government Act of 1959 paved the way for 'separate development', 'Bantu homelands' and 'multinational development'. It anticipated African home-lands gaining *separate* independence, possibly to form a commonwealth with white South Africa at its core. This vision of 'Grand Apartheid' became the ideal of Afrikaners: to them it was the only alternative to domination by the black majority. To achieve that ideal they had to divide the black majority. The key to that was land.

Perhaps apartheid began when Jan van Riebeeck, first governor of the Cape, planted a hedge of wild almonds to separate Dutch from Khoi-Khoi (Hottentots). Certainly from 1652 whites acquired increasingly more land: some was unoccupied, some was purchased from people who had no concept of buying and selling land because it was vested in the tribe and transactions in land were entirely outside their social experience and much was acquired by right of conquest. Africans were forced into the poorer parts of areas they had once occupied.

When the Union of South Africa was created in 1910 land for Africans was a major political issue. The Natives' Land Act of 1913 scheduled 22.5 million acres (9.1 million ha) as 'Native Reserves' for exclusive African occupation, but outside which no African could acquire land. The 'Scheduled Areas' amounted to 7.3 per cent of the area of South Africa at a time when Africans accounted for 67.3 per cent of the population. The inadequacy was evident, so the Beaumont Commission was set up to achieve a 'final settlement'. In 1916 it recommended the 'release' of a further 16.8 million acres (6.8 million ha) along

1913
Natives' Land Act:
scheduled areas

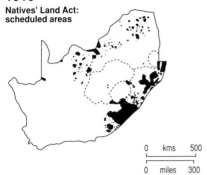

1916
Beaumont Commission
Proposals:
never implemented

0	kms	500
0	miles	300

1936
Native Trust and Land Act:
scheduled areas and
released areas

1975
Black Homelands:
actual black areas

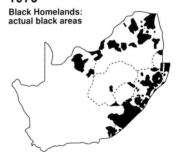

1975
Black Homelands:
consolidation proposals
not yet implemented

1992
Black Homelands:
'Independent'
homelands

Venda

Bophuthatswana

Ciskei

Transkei

with consolidation of the highly fragmented existing African lands. The resultant bill failed to emerge from parliamentary committee in 1917 and the proposals failed. The land question was settled, finally in Afrikaner eyes, in 1936 when the Native Trust and Land Act allowed for an additional 'quota' of 15.3 million acres (6.2 million ha) to be released for black occupation. The 1936 proposals have never been fully implemented.

Verwoerd's grand apartheid evolved from this basic division of land. The black areas were divided into ten 'homelands', four of which, Transkei (1976), Boputhatswana (1977), Venda (1979) and Ciskei (1982), were given an 'independence' not recognized beyond South Africa. The homelands emphasized tribalism with language as the main criterion of division but not rigorously so, as both Transkei and Ciskei are Xhosa speaking. The homelands are at best residual fragments of the land traditionally occupied by Africans.

The homelands are home to less than half of the blacks of South Africa. All blacks were made *de jure* citizens of a homeland even if they had been born and had resided all of their lives in white South Africa. The land area of the homelands is wholly inadequate even for the *de facto* population. Many homelands are fragments of land spattered across the map. Consolidation proposals tabled in 1975 aimed to reduce the number of parcels of land to thirty-four, much as Beaumont had proposed sixty years before.

The South African government embarked on one of the largest schemes of social/geographical engineering in the world. 'Black spots', areas occupied by black people in designated white areas, were systematically cleared and, with urban relocation and resettlement for consolidation and other purposes, over three million black people were forced to move. This forced resettlement in pursuit of the racial ideology of grand apartheid was the greatest single source of human misery and suffering in South Africa.

The homelands are residual economic backwaters away from major routes, minerals resources and the richest agricultural land. No large dam, power installation, irrigation scheme or factory sullies their fragmented faces. They are not economically viable but instead are dependent, much of their income coming from migrant and commuter workers employed in white South Africa. The South African government was never prepared to pay the price of achieving its own plans and consistently fell behind in implementing its own ideology, be it investment and job creation in the homelands following the Tomlinson Commission (1955), land acquisition for blacks following the Trust and Land Act (1936) or consolidation following Beaumont (1916). The commitment to the ideal of apartheid was less than wholehearted where it cost money.

The homelands were reservoirs of cheap labour essential for industrial white South Africa. They also provided a political façade to provide the white regime with legitimacy. But grand apartheid was a flawed concept because the ideal of racial separation could never be complete. White South Africa, always dependent on black labour, could never be otherwise.

The homelands are on the periphery of South Africa, but blacks could not be kept out of the urban areas, because their labour was needed in crucial industries and services. The Group Areas Act of 1955 and its many amendments controlled where people lived within South African towns but in confining urban blacks to designated townships and 'locations', there was a threat to white hegemony at the very heart of white South Africa. Urban blacks were essential to white prosperity but their very existence was the fatal flaw in the foundation of the whole structure of apartheid.

It was urban blacks who first drew the attention of the world to South Africa when, at a peaceful pass-law demonstration, 69 of them were shot dead at Sharpeville on 21 March 1960. It was urban blacks, mainly young people, who demonstrated against the Afrikaans medium in schools in Soweto in 1976 and were gunned down for their pains. It was urban blacks who voiced their objection to the constitutional reforms of 1984 and were shot in their hundreds before the world's television cameras. The South African police distinguished themselves by marking the twenty-fifth anniversary of Sharpeville by killing nineteen blacks at the Eastern Cape town of Uitenhage.

L'expérience apprend que le moment le plus dangereux pour un mauvais gouvernement est d'ordinaire celui où il commence à se réformer.

(Alexis de Tocqueville, 1856)

Grand Apartheid could not be reformed, yet the prosperity of white South Africa on which it depended could not survive without change.

The mantle of Verwoerd had been inherited by Balthazar Johannes Vorster in 1966 and his government carried through the ideals of Grand Apartheid, bringing the homelands to self-government and 'independence'. Through the 1970s capitalism and apartheid, which had for so long complemented and sustained each other, began to grow apart. South African economic success led business and industrial interests to want to loosen the ideological fetters of apartheid. Economic expansion demanded skilled labour and an end to job reservation whereby skilled jobs were kept for whites only. It demanded trade union reform and recognition of black trade unions to bring order to the labour market through formal negotiation and proper industrial relations. In June 1976 Soweto erupted when the South African police opened fire on children protesting against Afrikaans-medium teaching. The riots spread to other black townships and lasted well into 1977. The South African government conceded on Afrikaans-medium teaching but an official total of 575 blacks were killed and 2389 wounded in clashes with the police which were witnessed by the world's media. The brutal repression was roundly condemned and the first small but practical steps towards making South Africa a pariah state were taken: for example, in 1977 the Sullivan Code, a set of six principles designed to end racial segregation and job discrimination in American-owned companies in South Africa, and the Gleneagles Agreement by which British Commonwealth leaders agreed to discourage all sporting links with South Africa. Coinciding with this was Muldergate, or the Information scandal, when allegations of serious wrong-doing, dirty tricks, disinformation and non-accountability to parliament plunged South Africa into political crisis which brought down the Vorster government.

The National Party succession was won by P.W. Botha, but the problems confronting South Africa intensified and Botha limped from one reform of apartheid to another. Each was conceded grudgingly, each was too little, too late. Rather than helping they fuelled more protest, which was met with repression. Trade union reform (1979), job reservation reform (1980 and 1986), discontinuation of major resettlement (1985) were either too little, too late or both. After an all-white referendum (November 1983), a new constitution was introduced in September 1984. It gave, for the first time, representation to the Coloured and Indian population groups by means of a tri-cameral parliament but the reforms did not offer power-sharing and were deemed by the Coloured and Indian groups as largely cosmetic. Only 30.9 per cent and 20.3 per cent of the respective electorates even bothered to vote. Worse, there was no provision for Africans who comprised over two-thirds of the population. African protest was widespread. The 'reformist' government reverted to type and met protest with brutal repression. In July 1985 a state of emergency was declared in thirty-six districts and was extended nation-wide in June 1986. In the first half of 1986, on average five Africans per day were killed by the South African police. Despite strenuous efforts by the government to control the media, the outside world was made fully aware of the situation in South Africa and strongly disapproved. The piecemeal repeal of the Acts that had created the apartheid state, the Immorality Act (repealed 1985), the Mixed Marriages Act (repealed 1985), the pass laws (1986), and South African citizenship (1986), stopped neither internal dissension nor external approbation. International campaigns for sanctions against South Africa and for disinvestment gathered momentum. Barclays Bank, Ford, General Motors, IBM Computers and Kodak withdrew from South Africa, adding their prestigious names to over 200 other companies which pulled out of South Africa in the three years 1985–7 alone. Capitalist confidence in South Africa was severely undermined and the economy was in crisis, with high inflation, currency devaluation and high unemployment.

By the end of 1988 it became financially necessary for the South African government to negotiate withdrawal from Angola and Namibia. Botha had a stroke and was forced to resign. F.W. de Klerk, became State President in August 1989. Within six months he took radical steps which raised hopes of resolving the South African problem through negotiation. In early February 1990 the ANC was unbanned. The state of emergency was relaxed and Nelson Mandela was released after twenty-seven years in prison. Exiled members of the ANC were allowed to return to South Africa.

Not all parties are prepared to accept a cosy settlement between the National Party government and the ANC. To the right of the government are the white extremists, the Conservative Party and the AWB. Standing apart from the ANC is the Inkhata Freedom Party, led by Chief Buthelezi, which mainly represents Zulu interests. Inkhata has been largely responsible for the black on black violence in the townships, in its effort to obtain a greater say in any settlement. The government has gone along with this as divided blacks are likely to be less successful than united blacks. But when the white right-wing posed a threat through winning the whites-only by-election at Potchefstroom de Klerk countered with a whites-only referendum on 17 March 1992 in which he routed the right-wing, by 68.7 per cent to 31.3 per cent, on the question: 'Do you support continuation on the reform process which the State President began on 2 February 1990 and which is aimed at a new constitution through negotiation?'

Far from giving new impetus to the negotiations within the multi-party Convention for a Democratic South Africa (CODESA), by June 1992 they had stalled. Within the negotiations the stumbling block was minority rights and the precise percentage of votes required for constitutional change (an ironic reversal of the position of the National Party in the 1950s). The immediate cause of breakdown was township violence, in particular the Boipatong massacre by Inkhata supporters. It seems that the National Party, perhaps over-confident following the referendum, think that, with the collusion of some homeland leaders, they can prevent an ANC overall majority even with one person one vote. They wrongly anticipated the defeat of Mugabe in Zimbabwe and of Nujoma in Namibia, though the SWAPO failure to achieve a two-thirds majority encouraged them. But de Klerk would indeed be foolish to contemplate the defeat of Mandela in South Africa, though from de Klerk's point of view it would be equally foolish not to try the old divide-and-rule tactic. Another massacre, this time at Bisho by Ciskei homeland troops, prompted the resumption of negotiations in September 1992. There are time constraints on reaching a settlement: the whites-only election due in 1994 and the ageing of Nelson Mandela. Despite brinkmanship outside the arena and tortuous negotiations within, there is probably too much at stake for the main protagonists for there not to be a settlement in the allotted timescale, probably along the lines of majority rule with safeguards for minorities.

64 Southern Africa: the wind of change

In February 1960 the British Prime minister, Harold Macmillan, warned the South African parliament at Cape Town of a 'wind of change' sweeping through Africa which white South Africa would ignore at its peril. The subsequent events of 1960 in Africa were momentous. Sixteen new sovereign states came into being, the right of Africans to rule themselves conceded by Britain, France and Belgium. If Africans were fit to rule in Ghana, Nigeria and even Chad, could political rights continue to be denied to black South Africans? In South Africa itself 1960 witnessed the Sharpeville massacre, an attempt on Dr Verwoerd's life (albeit by a white man) and a state of emergency. But, before the year's end, white South Africa was greeting pitiful refugee convoys from the Congo (Zaire) almost with gratitude and a chorus of 'told you so'. In November white South Africans had sufficient confidence to support the 'miraculously' recovered Verwoerd by voting for a Republic and the following May cheered him on his return from London after taking South Africa out of the British Commonwealth. The state of emergency ended and foreign investment began to flow in again, all the more strongly, to help to create burgeoning economic prosperity, at least for whites. It slowly became clear that white South Africa was not going to heed warnings about the wind of change, that pursuit of the Afrikaner ideal of apartheid would continue with increased enthusiasm. Above all, the South African government with growing confidence sought actively to protect its apartheid state against outside interference particularly from elsewhere in Africa.

At the end of 1963 the short-lived Federation of Rhodesia and Nyasaland was broken up by the British Conservative government. The threat to South Africa seemed to increase when, in 1964, two component parts of the erstwhile Federation achieved their independence. Black Africa was now at the Zambesi, and even Algeria and Kenya, colonies with large white settler minorities, were independent states ruled by black majorities.

But the line of the Zambesi became South Africa's wind-break against the chill wind of change. South Africa ruled Namibia and with it the Caprivi strip, created in 1890 to give access to the Zambesi. At Katima Mulilo, where the strip reaches the Zambesi, South Africa created an advanced military base to defend the line. Further east the Zambesi formed the boundary between Zambia and white Rhodesia. The whole 550 mile (880 km) river and lake boundary was crossed by only two bridges, at the Victoria Falls and Chirundu, and by the road along the crest of the Kariba dam. They were well defended and it was not an easy frontier for guerillas to infiltrate against the South

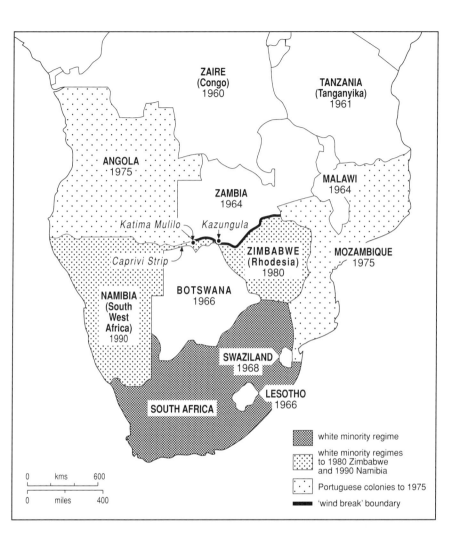

African and Rhodesian military. To east and west the wind-break frontier was buttressed by the Portuguese colonies of Mozambique and Angola, where liberation guerilla wars raged but posed no immediate threat to the white south.

The wind-break held for over ten years during which apartheid flourished. In November 1965 white Rhodesia unilaterally declared independence (UDI) from Britain and survived UN sanctions with the support of white South

Africa. Britain gave independence to Bechuanaland as Botswana and to Basutoland as Lesotho in 1966 and to Swaziland in 1968 but their formal independence was no more than an irritation to South African security. A renegotiated Southern African Customs Union (SACU), which came into effect in 1970, gave South Africa great economic control over the territories. There were tensions and cross-border clashes over the Zambesi, but South African policing of the line was very largely effective.

In 1975 the geo-politics of southern Africa was transformed when the two buttresses of the wind-break frontier, Angola and Mozambique, crumbled following the Portuguese *coup d'état* of 1974. South Africa's response was to invade Angola in October 1975 in an attempt to install UNITA, one of three rival Angolan liberation organizations, as the government in Luanda. The advance along the coast was halted by the intervention of the Soviet Union and Cuba and the South Africans fell back on defensive positions just inside Angola's southern boundary. The intervention was counter-productive because black African states rushed to recognize the MPLA as the Angolan government in November 1975. South African forces held their defensive line, which kept Namibia within the laager, with occasional sorties in support of UNITA and in 'hot pursuit' of SWAPO guerillas intent on penetrating into Namibia.

The right flank was more problematic. In Mozambique the liberation forces were united against the Portuguese and FRELIMO assumed power virtually unopposed. In March 1976 Mozambique closed all roads, railways and ports to Rhodesian traffic. They also opened a long land frontier of ideal guerilla country to the armed forces of Robert Mugabe's Zimbabwe African National Liberation Army (ZANLA). Despite building a railway direct to South Africa in October 1974, the days of white minority-ruled Rhodesia were numbered. An 'internal settlement' in 1978 failed to gain credibility or to stop guerilla incursions. Talks began with the British, and in December 1979 Rhodesian UDI was ended. Supervised elections led to the independence of Rhodesia as Zimbabwe in April 1980 under the radical leadership of Robert Mugabe. Within five years the frontier of black Africa had moved significantly forward. Mozambique and Zimbabwe each had long land frontiers with South Africa itself. Only in the west was the front-line kept remote from South Africa itself.

These events did not go unnoticed in South Africa. In June 1976 the Soweto uprising brought brutal repression and led to the most serious and widespread protest ever seen in South Africa. Although the primary trigger in Soweto was Afrikaans as a teaching medium, defeat of South African military forces in Angola and the independence of Mozambique were secondary causes. Under growing economic pressures white South Africa seemed to lose its confidence and sense of direction. Various commissions, on such matters as labour, trade unions and the constitution, attempted to reconcile apartheid with the changing economic and political scene.

The new Botha government was strongly influenced by military thinking and new strategies were developed to deal with the threat to apartheid that was posed by South Africa's neighbours. The concept of a constellation of

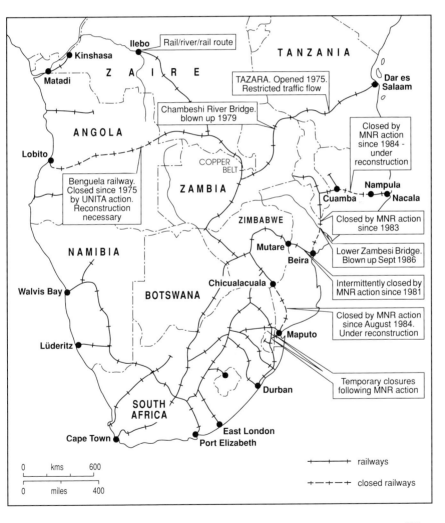

states was put forward in 1979, envisaging the neighbouring states in fixed orbit around a hegemonic South Africa. It was supported by a South African Development Bank tied to the homeland concept. The front-line states countered the constellation idea with the Southern Africa Development Co-ordination Conference (SADCC), which aimed to co-ordinate development, independent of South Africa, and to act as an umbrella to obtain aid from the industrialized countries

Realizing that its neighbours were not prepared to assume the dependent role it had envisaged for them, the South African government saw its apartheid state as under 'total onslaught'. Military thinking decreed that this was best countered by 'destabilization' of the neighbouring states through 'destructive engagement'.

In Angola civil war was fomented by direct South African support of UNITA with the backing of the United States. Angola was crippled by a war which forced the MPLA government to spend as much as 12 per cent of GNP on military spending. UNITA kept the Benguela railway closed and even invaded the diamond mines of the north-east. In Mozambique, the anti-FRELIMO Mozambique National Resistance (MNR), which had been formed with help from white Rhodesians, was now patronized by South Africa to enable it to wage a devastating civil war. This created more than a million refugees and starved tens of thousands to death. Mozambique's transport routes were systematically destroyed in order to increase the dependence of land-locked Zimbabwe and Zambia upon South Africa.

Swaziland was coaxed (hoaxed) into signing a secret non-aggression pact with South Africa in 1982 with the promise of a land-deal which never materialized. Mozambique was pressured into the similar, but much more publicized, Nkomati Accord in 1984. Under both agreements the ANC was excluded from each country in anything but a diplomatic role. Other neighbouring states resisted the signing of similar Accords with South Africa but were subjected to severe harassment. Military raids were mounted into Botswana, Lesotho, Swaziland, Zimbabwe and Zambia. The enclave state of Lesotho was blockaded until a military *coup d'état* ousted its government in January 1986. These actions were aimed at denying the ANC bases near South Africa and to make dependent hostages of the front-line states in order to deter the imposition of international economic sanctions against apartheid.

A rising tide of protest within, and international sanctions and disinvestment without, South Africa undermined its financial means and the resolve to fight on. Accommodation between the United States and the Soviet Union led to disengagement in Angola, and Namibia moved swiftly to independence in March 1990. That followed another change of government in South Africa which had led to the unbanning of the ANC and the release of Nelson

Mandela in February 1990. The hegemony of apartheid in southern Africa collapsed in true domino fashion. The future of southern Africa now lies in co-operation between all states with perhaps a majority-ruled South Africa taking a leading role in co-ordinated development.

65 Southern Africa: challenge of the future

The 1980s was the decade of destabilization and destructive engagement in southern Africa, during which the largest and richest state, South Africa, did all that it could to undermine the well-being of its neighbours in an effort to preserve apartheid. The 1990s could be the decade for co-operation between all of the states of southern Africa in an effort to deal with the basic problems that beset the sub-continental region. Even without the wars generated by the death throes of apartheid (and they splutter on), problems in the region abound: drought, which in 1992 caused economic crisis in Zambia, Zimbabwe, Botswana and South Africa; typhoons in Mozambique; accelerating population growth; AIDS; land alienation; unemployment; accumulated debt; economic diversification; regional inequalities; urbanization and inadequate housing; badly maintained railways; and under-financed ports.

The confrontation of the 1980s threw up appropriate structures: on the one hand SADCC, on the other the constellation of states. The latter slowly withered away but its shadow, that of South African economic and political hegemony, remains. A black majority-ruled South Africa will be no less economically dominant within the region than was the apartheid state. In rough terms South Africa represents 20 per cent of the land area, 40 per cent of the population and 80 per cent of the wealth of the whole sub-continental region, leaving the other nine states with the reciprocal percentages. It spells out some of the enormous regional inequalities and hides others as, for example, between the different parts of South Africa itself. Even more important is the fact that, although it has not been easy for the front-line states to resist economic domination by a hostile South Africa, it is not going to be easy for them to resist domination by a friendly South Africa. True, they will not have to contend with sabotage and destruction of transport facilities but co-operation with such a dominant economic force also has its problems.

Leaders of the SADCC states have been quick to express the delight that they would have in welcoming a majority-ruled South Africa into membership. This is understandable but in direct contradiction of one of the original basic aims of SADCC: 'to reduce the historic dependence of these countries [SADCC] on South Africa by delinking them from South Africa'. Can SADCC accommodate such a fundamental change? A cynical view may be that so little has been achieved that there is no problem, even if SADCC were to have to stand on its head.

There is a problem, however, and all the more so if the SADCC achieve-

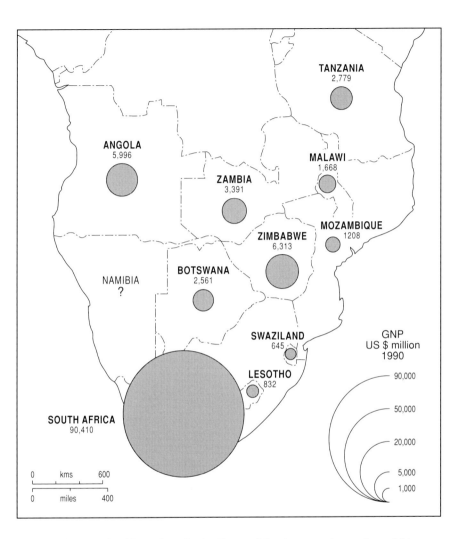

TANZANIA
2,779

ANGOLA
5,996

MALAWI
1,668

ZAMBIA
3,391

MOZAMBIQUE
1208

ZIMBABWE
6,313

NAMIBIA
?

BOTSWANA
2,561

SWAZILAND
645

GNP
US $ million
1990

LESOTHO
832

90,000

50,000

SOUTH AFRICA
90,410

20,000

5,000

1,000

0   kms   600

0   miles   400

ments have indeed been few. As the focus of development in southern African moves beyond transportation towards agriculture and industrialization, the relative size, depth of infrastructure and potential strength of a South African economy unburdened by large military spending will be hard to resist. A radical re-think of SADCC structures is necessary, perhaps by building in a positive political dimension rather than gingerly treading around the fringes of economic co-operation, which has been characteristic of SADCC thus far. The challenge ahead for southern Africa is enormous and fundamental questions have to be tackled before any significant progress can be made.

66 Further reading

Amin S., Chitala, D. and Mandaza, I (eds) (1987) *SADCC: Prospects for Disengagement and Development in Southern Africa*, London: ZED.

Brittain, V. (1988) *Hidden Lives, Hidden Deaths: South Africa's Crippling of a Continent*, London: Faber & Faber.

Commonwealth Eminent Persons Group (1986) *Mission to South Africa: the Commonwealth Report*, London: Penguin.

Commonwealth Expert Group (1991) *Beyond Apartheid: Human Resources in a New South Africa*, London: James Currey.

Geldenhuys, D. (1984) *The Diplomacy of Isolation: South African Foreign Policy Making*, Johannesburg: Macmillan.

Giliomee, H. and Schlemmer, L. (1990) *From Apartheid to Nation-Building*, Cape Town: Oxford University Press.

Griffiths, I. Ll. and Funnell, D.C. (1991) 'The abortive Swazi land-deal', *African Affairs* 90 (358): 51–64.

Hanlon, J. (1986) *Apartheid's Second Front*, London: Penguin.

Johnston, R.W. (1977) *How Long will South Africa Survive?*, London: Macmillan.

Kgarebe, A. (1981) *SADCC 2: Maputo*, London: SADCC Liaison Committee.

Lemon, A. (1987) *Apartheid in Transition*, Aldershot: Gower.

Libby, R.T. (1987) *The Politics of Economic Power in Southern Africa*, Princeton, NJ: Princeton University Press.

Omer-Cooper, J.D. (1987) *History of Southern Africa*, London: James Currey.

Platsky, L. and Walker, C. (1985) *The Surplus People: Forced Removals in South Africa*, Johannesburg: Ravan.

Rogers, B. (1980) *Divide and Rule: South Africa's Bantustans*, London: International Defence and Aid Fund.

Sampson, A. (1987) *Black and Gold: Tycoons, Revolutionaries and Apartheid*, London: Hodder & Stoughton.

Sparks, A. (1990) *The Mind of South Africa: The Story of the Rise and Fall of Apartheid*, London: Heinemann.

Smith, S. (1990) *Front Line Africa*, Oxford: OXFAM.

Thompson, L. (1985) *The Political Mythology of Apartheid*, New Haven, Conn: Yale University Press.

Appendices

67 Chronology of African independence

State	Date of independence	Colonial power	Notes
Ethiopia	Ancient	–	Italian occupation 1936–41.
Liberia	26.7.1847	–	Private colony 1822–47. Home for freed American slaves.

South Africa	31.5.1910	Britain	(*Suid Afrika*) Union of four colonies, Cape Colony, Natal, Orange River Colony (*Oranje Vrij Staat*) and Transvaal (*Zuid Afrikaansche Republiek*), the last two of which had been independent Boer republics to 31.5.1902. The Union became republic outside British Commonwealth 31.5.1961. White minority rule. Unrecognized 'independent' homelands:
			  Transkei 26.10.1976
			  Boputhatswana 6.12.1977
			  Venda 13.9.1979
			  Ciskei 4.12.1981

Egypt	28.2.1922	Britain	Joined with Syria as United Arab Republic (UAR) from 1.2.1958 to 28.9.1961. Federated with Kingdom of (North) Yemen from 8.3.1958 to 26.12.1961. Name UAR retained by Egypt to 2.9.1971.
Libya	24.12.1951	Italy	British (Tripolitania and Cyrenaica) and French (Fezzan) administration 1943–51.

State	Date of independence	Colonial power	Notes
Ethiopia (Ogaden)	28.2.1955		Italian occupation 1936–41. British administration 1941–55.
Sudan	1.1.1956	Britain & Egypt	Anglo-Egyptian condominium.
Morocco	2.3.1956	France	(*Maroc*)
Tunisia	20.3.1956	France	(*Tunisie*)
Morocco (part)	7.4.1956	Spain	(*Marruecos*) Spanish northern zone.
Morocco (part)	29.10.1956		International zone (Tangiers).
Ghana	6.3.1957	Britain	(Gold Coast) including British Togoland (UN Trust), part of former German colony of Togo.
Morocco (part)	27.4.1958	Spain	(*Marruecos*) Spanish southern zone.
Guinea	2.10.1958	France	(*Guinée Française*)
Cameroon	1.1.1960	France	(*Cameroun*) UN Trust. Larger part of former German colony of *Kamerun*.
Togo	27.4.1960	France	UN Trust. Larger part of former German colony of Togo.
Senegal	20.6.1960 (20.8.1960)	France	First independent as 'Federation of Mali' with Mali (former French Soudan). Federation dissolved after two months. Joined Gambia in Confederation of Senegambia, 1.1.1982 to 6.10.1989.
Mali	20.6.1960 (22.9.1960)	France	(*Soudan Française*) Independent initially as 'Federation of Mali' with Senegal. Federation dissolved after two months.
Madagascar	26.6.1960	France	(Malagasy, *République Malagache*)

State	Date of independence	Colonial power	Notes
Zaire	30.6.1960	Belgium	Congo Free State (*Etat Indépendant du Congo*) 2.5.1885 to 18.11.1908 when it became the Belgian Congo (*Congo Belge, Belgisch Congo*). Name changed from Congo 27.10.1971.
Somalia	1.7.1960	Italy Britain	UN Trust. Union of two colonies, Italian and British Somaliland. British Somaliland independent prior to union on 26.6.1960.
Benin	1.8.1960	France	Name changed from Dahomey 30.11.1975.
Niger	3.8.1960	France	
Burkina Faso	5.8.1960	France	Name changed from Upper Volta (*Haute Volta*) 4.8.1984.
Ivory Coast	7.8.1960	France	(*Côte d'Ivoire*)
Chad	11.8.1960	France	(*Tchad*)
Central African Republic (CAR)	13.8.1960	France	(*Obangui-Chari, République Centrafricaine*) Central African Empire from 4.12.1976 to 20.9.1979.
Congo (Brazzaville)	15.8.1960	France	(*Moyen Congo*)
Gabon	17.8.1960	France	
Nigeria	1.10.1960	Britain	
Mauritania	28.11.1960	France	(*Mauritanie*)
Sierra Leone	24.4.1961	Britain	
Nigeria (British North Cameroon)	1.6.1961	Britain	UN Trust. Part of former German colony of *Kamerun*. Plebiscite 11/12.2.1961.
Cameroon (British South Cameroon)	1.10.1961	Britain	UN Trust. Part of former German colony of *Kamerun*. Plebiscite 11/12.2.1961. Union with Cameroon as United Republic of Cameroon.

State	Date of independence	Colonial power	Notes
Tanzania	9.12.1961	Britain	(Tanganyika) UN Trust. Greater part of former German colony of *Deutsche Ostafrika*. Name changed to Tanzania following union with Zanzibar 27.4.1964.
Burundi	1.7.1962	Belgium	UN Trust. Ruanda–Urundi, divided at independence, was smaller part of former German colony of *Deutsche Ostafrika*.
Rwanda	1.7.1962	Belgium	
Algeria	3.7.1962	France	(*Algérie*)
Uganda	9.10.1962	Britain	
Tanzania (Zanzibar)	10.12.1963	Britain	Union with Tanganyika as Tanzania 27.4.1964.
Kenya	12.12.1963	Britain	
Malawi	6.7.1964	Britain	(Nyasaland) Federated with Rhodesias 1.10.1953 to 31.12.1963.
Zambia	25.10.1964	Britain	(Northern Rhodesia) Federated with Nyasaland and Southern Rhodesia 1.10.1953 to 31.12.1963.
Gambia	18.2.1965	Britain	Joined with Senegal as Confederation of Senegambia, 1.1.1982 to 6.10.1989.
Botswana	30.9.1966	Britain	(Bechuanaland)
Lesotho	4.10.1966	Britain	(Basutoland)
Mauritius	12.3.1968	Britain	
Swaziland	6.9.1968	Britain	
Equatorial Guinea	12.10.1968	Spain	Comprises *Rio Muni* and *Macias Nguema Biyogo* (*Fernando Poo*).
Morocco (Ifni)	30.6.1969	Spain	(*Territorio de Ifni*)
Guinea-Bissau	10.9.1974	Portugal	*Guine-Bissau* formerly *Guine-Portuguesa*.
Mozambique	25.6.1975	Portugal	(*Moçambique*)

State	Date of independence	Colonial power	Notes
Cape Verde	5.7.1975	Portugal	(*Cabo Verde*)
Comoros	6.7.1975	France	*Archipel des Comores*. Excluding island of Mayotte which remains a French Overseas Territory (*Territoire d'Outre-Mer*).
St Thomas and Prince Islands	12.7.1975	Portugal	(*São Tomé e Príncipe*)
Angola	11.11.1975	Portugal	Includes detached enclave of Cabinda.
Western Sahara	28.2.1976	Spain	(*Rio de Oro* and Sequit el Hamra) On Spanish withdrawal seized by Morocco. Occupation disrupted by POLISARIO, formed 10.5.1973.
Seychelles	26.6.1976	Britain	
Djibouti	27.6.1977	France	(*Territoire Française des Afars et des Issas* formerly *Côte Française des Somalis*)
Zimbabwe	18.4.1980	Britain	(Rhodesia, formerly Southern Rhodesia) Unilateral Declaration of Independence (UDI) in effect from 11.11.1965 to 12.12.1979. Federated with Northern Rhodesia and Nyasaland 1.10.1953 to 31.12.1963.
Namibia	21.3.1990	South Africa	(South West Africa) UN Trust. Former German colony of *Deutsche Südwesafrika*.
Eritrea	24.5.1993	Italy Ethiopia	British administration 1941–52. Federated with Ethiopia 11.9.1952. Union with Ethiopia 14.11.1962.

State	Date of independence	Colonial power	Notes
African territories and islands not independent			
Spanish North Africa		Spain	*Plazas de Soberania: Ceuta, Islas Chafarinas, Melilla, Penon de Velez de la Gomera, Penon de Alhucemas.* Small enclaves and islands on the north coast of Morocco.
Madeira		Portugal	(*Arquipelago da Madeira*)
Canary Islands		Spain	(*Islas Canarias*)
St Helena with Ascension and Tristan da Cunha		Britain	British Crown Colony.
Socotra		Yemen	
Mayotte		France	Island of Comoros Group. *Territoire Française d'Outre-Mer.*
Reunion		France	*Ile de la Réunion, Département d'Outre-Mer* (from 1946).
French Indian Ocean Islands		France	*Ile Europa, Ile Juan de Nova, Bassas da India, Iles Glorieuses, Ile Tromelin* (All near Madagascar).

68 States, capitals, changes of government and political leaders in post-colonial Africa

State	Capital	(P/L)	Dates A	B	C	Leaders
Algeria	Algiers	P	1962			Ben Bella
					1965	*Boumedienne* (d)
				1979		*Chadli Benjedid*
					1992	Boudiaf (a)
				1992		
Angola	Luanda	P	1975			Neto (d)
				1979		Dos Santos
Benin	Porto Novo	P	1960			Maga
					1963	Ahomadegbe/Apithy
					1965	*Soglo*
					1967	*Alley*
					1969	*De Souza*
					1972	*Kerekou*
Botswana	Gaborone	L	1966			Khama (d)
				1980		Masire
Burkina Faso	Ouagadougou	L	1960			Yameogo
					1966	*Lamizana*
					1980	*Zerbo*
					1982	*Ouedraogo*
					1983	*Sankara* (a)
					1987	*Campaore*
Burundi	Bujumbura	L	1962			(Mwami) Mwambutse IV
					1966	(Mwami) Ntare V (e 1972)
					1966	*Micombero*
					1976	*Bagaza*
					1987	*Buyoya*
Cameroon	Yaoundé	–	1960			Ahidjo
				1982		Biya
Cape Verde	Praia	P	1975			Pereira
				1991		Monteiro

State	Capital	(P/L)	Dates			Leaders
			A	B	C	
Central African Republic	Bangui	L	1960			Dacko
					1965	*Bokassa*
					1979	Dacko
					1981	*Kolingba*
Chad	N'Djamena	L	1960			Tombalbaye (a)
					1975	*Malloum*
					1979	*Oueddei*
					1982	*Habré*
					1990	*Deby*
Comoros	Moroni	P	1975			Solih
					1978	Abdallah (a)
					1990	Taki
Congo	Brazzaville	–	1960			Youlou
					1963	Massamba-Debat (e)
					1968	*Ngouabi* (a)
					1977	*Yhombi-Opango*
				1979		*Sassou-Nguesso*
Djibouti	Djibouti	P	1977			Gouled
Egypt	Cairo	–	1922			
				1936		(King) Farouk
					1952	*Neguib*
					1954	*Nasser* (d)
				1970		*Sadat* (a)
				1981		*Mubarak*
Equatorial Guinea	Malabo	P	1968			Nguema (e)
					1979	*Mbasogo*
Eritrea	Asmara	–	1993			Afeworki
Ethiopia	Addis Ababa	–	Ancient			
				1930		(Emperor) Haile Selassie
					1974	*Aman Adom* (a)
				1974		*Teferi Banti* (e)
				1977		*Mengistu*
					1991	Zenawi
Gabon	Libreville	P	1960			M'Ba (d)
				1967		Bongo
Gambia	Banjul	P	1965			Jawara

State	Capital	(P/L)	Dates A	B	C	Leaders
Ghana	Accra	P	1957			Nkrumah
					1966	*Ankrah*
				1969		*Afrifa* (e 1979)
				1969		Busia
					1972	*Acheampong* (e 1979)
					1978	*Akuffo* (e 1979)
					1979	*Rawlings*
				1979		Limann
					1981	*Rawlings*
Guinea	Conakry	P	1958			Sekou Toure (d)
				1984		Beavogui
					1984	*Conte*
Guinea-Bissau	Bissau	P	1974			Luiz Cabral
					1980	*Vieira*
Ivory Coast	Yamoussoukro	–	1960			Houphouët-Boigny
Kenya	Nairobi	–	1963			Kenyatta (d)
				1978		Arap-Moi
Lesotho	Maseru	L	1966			(King) Moshoeshoe II
					1970	Jonathan
					1986	*Lekhanya*
					1991	*Ramaema*
Liberia	Monrovia	P	1847			
				1944		Tubman (d)
				1971		Tolbert (a)
					1980	*Doe* (a)
					1990	No clear leader emerged
Libya	Tripoli	P	1951			(King) Idris
					1969	*Gadafy*
Madagascar	Antananarivo	–	1960			Tsiranana
					1972	*Ramanantsoa*
				1975		*Ratsimandrava* (a)
				1975		*Anriamahazo*
				1975		*Ratsiraka*
Malawi	Lilongwe	L	1964			Banda
Mali	Bamako	L	1960			Keita
					1968	*Traore*

State	Capital	(P/L)	Dates A	B	C	Leaders
Mauritania	Nouakchott	P	1960			Ould Daddah
				1978		*Ould Salek*
				1979		*Ould Boucief* (d)
				1979		*Heydallah*
				1984		*Ould Taya*
Mauritius	Port Louis	P	1968			Ramgoolam
				1982		Jugnauth
Morocco	Rabat	P	1956			(King) Mohammed V (d)
				1961		(King) Hassan II
Mozambique	Maputo	P	1975			Machel (d)
				1986		Chissano
Namibia	Windhoek	–	1990			Nujoma
Niger	Niamey	L	1960			Diori
				1974		*Kountche* (d)
				1987		*Saibou*
Nigeria	Lagos	P	1960			Balewa (a)
				1966		*Ironsi*
				1966		*Gowon*
				1975		*Murtala Mohammed* (a)
				1976		*Obasanjo*
				1979		Shehu Shagari
				1983		*Buhari*
				1985		*Babangida*
Rwanda	Kigali	L	1962			Kayibanda
				1973		*Habyarimana*
St Thomas & Prince Islands	São Tomé	P	1975			Da Costa
Senegal	Dakar	P	1960			Senghor
				1981		Diouf
Seychelles	Victoria	P	1976			Mancham
				1977		René
Sierra Leone	Freetown	P	1961			Milton Margai (d)
				1964		Albert Margai
				1967		*Juxon-Smith*
				1968		Stevens
				1985		*Momoh*

State	Capital	(P/L)	Dates A	B	C	Leaders
Somalia	Mogadishu	P	1960			Shirmarke (a 1969)
				1964		Haji Hussein
				1967		Egal
					1969	*Siad Barre*
					1991	*Ali Mahdi Mohamed*
South Africa	Pretoria	–	1910			
				1948		Malan
				1954		Strijdom (d)
				1958		Verwoerd (a)
				1966		Vorster
				1978		Botha
				1989		De Klerk
Sudan	Khartoum	–	1956			Khalil
					1958	*Abboud*
					1964	*al-Khalifa*
				1965		Mahgoub
				1966		Sadiq al-Mahdi
					1969	*Nimeiri*
					1985	*Swar el-Dahab*
				1986		Sadiq al-Mahdi
					1988	*al-Bashir*
Swaziland	Mbabane	L	1968			(King) Sobhuza II (d)
				1982		(Regency)
				1986		(King) Mswati III
Tanzania	Dar es Salaam	P	1961			Nyerere
				1985		Mwinyi
Togo	Lomé	P	1960			Olympio (a)
					1963	Grunitzky
					1967	*Eyadema*
Tunisia	Tunis	P	1956			(Bey) Sidi Mohammed al-Amin/Borguiba
					1957	Borguiba
					1988	*Ben Ali*

State	Capital	(P/L)	Dates A	B	C	Leaders
Uganda	Kampala	L	1962			(Kabaka) Mutesa II/Obote
					1966	Obote
					1971	*Amin*
					1979	Lule
				1979		Binaisa
				1980		Obote
					1985	*Okello*
					1986	*Museveni*
Zaire	Kinshasa	–	1960			Lumumba (e 1961)
					1960	*Mobutu*
				1961		Adoula
				1964		Tshombe
					1965	*Mobutu*
Zambia	Lusaka	L	1964			Kaunda
				1991		Chiluba
Zanzibar*	Zanzibar	P	1963			(Sultan) bin Khalifa
					1964	Karume (a 1972)
Zimbabwe	Harare	L	1980			Mugabe

Notes: CAPITAL P – port capital city [28], L – land-locked state [14]

DATES A – independence [52], B – constitutional change of government [48], C – successful *coup d'état* [80]

LEADERS (a) – assassinated [16], (d) – died in office [15], (e) – executed [8] Military leaders in italics [75]

* Merged with Tanganyika to form Tanzania 27.4.1964.

Successful *coups d'état*

1950	1960 *	1970 *	1980 ***	1990 ***
1951	1961	1971 *	1981 **	1991 ***
1952 *	1962	1972 ***	1982 **	1992 *
1953	1963 ***	1973 *	1983 **	1993
1954 *	1964 **	1974 **	1984 **	
1955	1965 ****	1975 **	1985 ***	
1956	1966 *******	1976 *	1986 **	
1957 *	1967 ***	1977 **	1987 **	
1958 *	1968 ***	1978 ***	1988 **	
1959	1969 ****	1979 ******	1989	Total 80

Constitutional changes of government

1950	1960	1970 *	1980 **	1990
1951	1961 **	1971 *	1981 **	1991 **
1952	1962	1972	1982 ***	1992
1953	1963	1973	1983	1993
1954 *	1964 ***	1974 *	1984 *	
1955	1965 *	1975 ***	1985 **	
1956	1966 **	1976 *	1986 ***	
1957	1967 **	1977 *	1987 *	
1958 *	1968	1978 **	1988	
1959	1969 **	1979 *******	1989 *	Total 48

Military leaders in power (military leaders/no. of independent states: percentage)

1950	1/4	25	1960	3/27	11	1970	14/42	33	1980	21/51	41	1990	29/52	56
1951	1/5	20	1961	3/29	10	1971	15/42	36	1981	23/51	45	1991	29/52	56
1952	1/5	20	1962	2/33	6	1972	17/42	40	1982	23/51	45	1992	27/52	52
1953	1/5	20	1963	2/35	6	1973	18/42	43	1983	24/51	47	1993		
1954	1/5	20	1964	2/36	6	1974	20/43	47	1984	25/51	49			
1955	1/5	20	1965	6/37	16	1975	21/48	44	1985	27/51	53			
1956	1/8	13	1966	9/39	23	1976	21/49	43	1986	28/51	55			
1957	1/9	11	1967	11/39	28	1977	21/50	42	1987	28/51	55			
1958	2/10	20	1968	13/42	31	1978	22/50	44	1988	29/51	57			
1959	2/10	20	1969	15/42	36	1979	23/50	46	1989	29/51	57			

69 Area, population and gross national product (GNP); population density, GNP per caput, growth rate GNP per caput

Country	Area (sq. km)	GNP (million US $) 1991	Pop ('000 1990)	Persons (per sq. km 1990)	GNP per caput US $ 1991	Growth rate GNP per caput 1980–91
Algeria	2,381,741	52,239	25,056	10.5	2020	−0.8
Angola	1,246,700	5,996*	10,011	8.0	620*	6.1**
Benin	112,622	1,848	4,741	42.1	380	−1.1
Botswana	600,372	3,335	1,254	2.1	2590	5.8
Burkina Faso	274,200	3,213	9,016	32.9	350	1.3
Burundi	27,834	1,210	5,470	196.5	210	1.4
Cameroon	475,442	11,320	11,941	25.1	940	−0.9
Cape Verde	4,033	285	371	92.0	750	2.2
Central African R.	622,984	1,218	3,036	4.9	390	−1.5
Chad	1,284,000	1,212	5,679	4.4	220	3.8
Comoros	2,400	245	475	197.9	500	−1.0
Congo	342,000	2,623	2,277	6.7	1120	−0.2
Djibouti	22,733	–	427	18.8	–	–
Egypt	1,001,449	33,068	52,061	52.0	620	2.0
Equatorial Guinea	28,051	142	417	14.9	330	3.4
Ethiopia	1,221,900	6,144	51,183	41.9	120	−1.6
Gabon	267,667	4,419	1,135	4.2	3780	−4.2
Gambia	11,295	322	875	77.5	360	−0.1
Ghana	238,537	6,176	14,870	62.3	400	−0.3
Guinea	254,857	2,669	5,718	22.4	450	–
Guinea-Bissau	36,125	194	981	27.2	190	1.3
Ivory Coast	322,463	8,523	12,233	37.9	690	−3.4
Kenya	582,645	8,505	24,368	41.8	340	0.3
Lesotho	30,355	1,053	1,771	58.3	580	0.0
Liberia	111,369	–	2,560	23.0	–	–
Libya	1,759,540	23,333*	4,546	2.6	5310*	−9.2**
Madagascar	687,041	2,560	11,620	16.9	210	−2.4
Malawi	118,484	1,996	8,504	71.8	230	0.1

Country	Area (sq. km)	GNP (million US $) 1991	Pop ('000 1990)	Persons (per sq. km 1990)	GNP per caput US $ 1991	Growth rate GNP per caput 1980—91
Mali	1,240,000	2,412	8,461	6.8	280	−0.1
Mauritania	1,030,700	1,026	1,969	1.9	510	−1.8
Mauritius	2,045	2,623	1,074	525.2	2420	6.1
Morocco	446,550	26,451	25,091	56.2	1030	1.6
Mozambique	783,030	1,163	15,784	20.2	70	−3.6
Namibia	824,292	–	1,780	–	–	−1.5
Niger	1,267,000	2,361	7,666	6.1	300	−4.1
Nigeria	923,768	34,057	117,510	127.2	290	−1.7
Rwanda	26,338	1,930	7,113	270.1	260	−2.6
St Thomas & Prince	964	42	123	127.6	350	−3.5
Senegal	196,192	5,500	7,428	37.9	720	0.0
Seychelles	518	350	68	131.3	5110	2.5
Sierra Leone	71,740	904	4,137	57.7	210	−1.3
Somalia	637,657	946**	6,284	9.9	150**	−1.8**
South Africa	1,221,037	90,953	35,914	29.4	2530**	0.9
Sudan	2,505,813	10,107	25,191	10.1	400**	−2.4
Swaziland	17,363	874	789	45.4	1060	3.1
Tanzania	945,087	2,424	24,518	25.9	100	−1.1
Togo	56,000	1,530	3,638	65.0	410	−1.7
Tunisia	164,150	12,417	8,175	49.8	1510	1.2
Uganda	236,036	2,762	17,358	73.5	160	3.3
Zaire	2,354,409	8,123	35,564	15.1	220**	−1.6
Zambia	752,614	3,394	8,122	10.8	420**	−2.9
Zimbabwe	390,580	6,220	9,531	24.4	620	0.2
Total/Average	30,162,722	–	645,884	21.4	–	–

Source: IBRD (1991; 1992) *World Bank Atlas*
Notes: * 1989 figures.
    ** 1990 figures.
    – Data not available.
    Statistics in these tables should be treated with caution. Although from the most authoritative and most readily available sources, they are at best approximations and are subject to retrospective revision.

Other African territories

Territory	Area (sq. km)
Canary Islands	7,273
French Indian Ocean Islands	50
Madeira	796
Mayotte	374
Reunion	2,510
St Helena and dependencies	396
Socotra	3,626
Spanish North Africa	36
Western Sahara	266,770
Total	30,444,553

70 Military expenditure (per cent of GNP), arms imports (million US $), debt (per cent of GNP), debt servicing (per cent of exports), workers' remittances from abroad (per cent of GNP), official development assistance (ODA) (per cent of GNP), balance of payments (million US $)

	Military expend. % GNP 1989	Arms imports US $ m. 1989	Debt % GNP 1989	Debt service % exports 1988	Workers' rems % GNP 1989	ODA % GNP 1989	Balance pay. US $ m. 1990
Algeria	1.9	930	57	77	0.7	0.3	−2040
Angola	21.5	3592	–	–	–	–	367
Benin	1.9	13	72	5	3.3	16.1	−177
Botswana	1.9	76	23	4	–	8.4	309
Burkina Faso	2.8	3	30	12	5.4	14.5	−310
Burundi	2.6	20	82	25	–	16.5	−163
Cameroon	2.1	17	44	12	0.0	4.2	−881
Cape Verde	–	–	–	–	–	–	–
Central African R.	1.7	6	66	6	−2.5	17.5	−181
Chad	3.8	100	37	3	−2.0	28.1	−252
Comoros	–	–	81	23	–	21.5	–
Congo	3.6	2	215	29	−2.7	4.7	7
Djibouti	–	–	–	7	–	–	–
Egypt	4.5	4717	159	14	13.1	4.7	−2848
Equatorial Guinea	–	–	146	23	–	30.0	–
Ethiopia	13.6	629	51	37	–	12.2	−510
Gabon	4.5	143	102	6	−4.9	4.2	−627
Gambia	–	–	162	17	–	52.2	–
Ghana	0.6	37	60	20	0.1	9.7	−232

	Military expend. % GNP 1989	Arms imports US$m. 1989	Debt % GNP 1989	Debt service % exports 1988	Workers' rems % GNP 1989	ODA % GNP 1989	Balance pay. US$m. 1990
Guinea	3.0	85	85	22	–	15.0	−279
Guinea-Bissau	–	–	–	31	–	–	–
Ivory Coast	1.2	51	182	13	−5.5	4.8	−1335
Kenya	2.6	185	72	19	0.0	11.6	−711
Lesotho	2.4	5	39	5	–	17.1	−130
Liberia	2.2	11	–	–	–	5.9	−163
Libya	7.4	2247	–	–	−2.2	0.0	−2222
Madagascar	1.3	30	154	39	–	15.4	−261
Malawi	1.6	23	91	17	–	29.8	−134
Mali	3.3	26	105	14	1.8	26.1	−350
Mauritania	5.7	0	213	22	0.4	21.4	−179
Mauritius	0.3	22	41	10	0.0	3.0	−65
Morocco	4.3	510	98	25	6.0	2.5	164
Mozambique	10.4	19	427	8	–	49.0	−733
Namibia	–	–	–	–	–	2.9	–
Niger	0.8	5	79	21	−1.8	13.5	−248
Nigeria	1.0	432	119	24	−0.1	1.1	−1045
Rwanda	1.7	7	30	10	−0.8	11.5	−258
St Thomas & Prince	–	–	–	16	–	–	–
Senegal	2.0	30	93	18	0.6	14.4	−467
Seychelles	–	–	51	–	–	7.3	–
Sierra Leone	0.5	9	–	6	0.0	10.6	−86
Somalia	3.0	43	203	5	–	45.4	−349
South Africa	4.2	208	–	–	–	–	–
Sudan	5.9	189	71	10	3.7	9.4	−1144
Swaziland	–	–	45	–	–	5.0	–
Tanzania	5.2	110	186	17	–	24.3	−743
Togo	3.2	74	91	18	0.3	14.7	−122
Tunisia	4.9	70	72	24	4.8	2.6	93
Uganda	4.2	33	39	14	–	8.9	−289
Zaire	1.2	28	97	7	0.0	11.1	−888
Zambia	3.2	0	159	14	−0.7	18.0	−234
Zimbabwe	7.9	284	54	25	0.0	4.4	−56
Totals		15021 (42)				(41)	−19772

Source: UNDP (1902: 1992) *Human Development Report*

71 Human development index (HDI), HDI rank, GNP per caput rank, life expectancy at birth 1990, adult literacy 1990

	HDI	HDI rank	GNP per caput rank	HDI rank minus GNP rank	Life expect. 1990	Adult literacy (%) 1990
Algeria	.533	8	7	− 1	65.1	57.4
Angola	.169	34	16	−18	45.5	41.7
Benin	.111	42	28	−14	47.0	23.4
Botswana	.534	7	4	− 3	59.8	73.6
Burkina Faso	.074	50	30	−20	48.2	18.2
Burundi	.165	36	41	+ 5	48.5	50.0
Cameroon	.313	20	12	− 8	53.7	54.1
Cape Verde	.437	10	13	+ 3	67.0	53.0
Central African R.	.159	38	27	−11	49.5	37.7
Chad	.088	43	39	− 4	46.5	29.8
Comoros	.269	24	21	− 3	55.0	61.0
Congo	.372	16	9	− 7	53.7	56.6
Djibouti	.084	46	–		48.0	19.0
Egypt	.385	14	16	+ 2	60.3	48.4
Equatorial Guinea	.163	37	33	− 4	47.0	50.2
Ethiopia	.173	33	46	+13	45.5	66.0
Gabon	.545	6	3	− 3	52.5	60.7
Gambia	.083	47	29	−18	44.0	27.2
Ghana	.310	21	25	+ 4	55.0	60.3
Guinea	.052	52	22	−30	43.5	24.0
Guinea-Bissau	.088	45	44	− 1	42.5	36.5
Ivory Coast	.289	23	15	− 8	53.4	53.8
Kenya	.366	17	32	+ 15	59.7	69.0
Lesotho	.423	12	19	+ 7	57.3	78.0
Liberia	.227	28	–		54.2	39.5
Libya	.659	4	1	− 3	61.8	63.8
Madagascar	.325	18	41	+23	54.5	80.2
Malawi	.166	35	38	+ 3	48.1	47.0
Mali	.081	48	36	−12	45.0	32.0
Mauritania	.141	41	20	−19	47.0	34.0

	HDI	HDI rank	GNP per caput rank	HDI rank minus GNP rank	Life expect. 1990	Adult literacy (%) 1990
Mauritius	.793	I	6	+ 5	69.6	86.0
Morocco	.429	II	II	0	62.0	49.5
Mozambique	.153	40	49	+ 9	47.5	32.9
Namibia	.295	22	–		57.5	40.0
Niger	.078	49	34	−15	45.5	28.4
Nigeria	.241	27	35	+ 8	51.4	50.7
Rwanda	.186	31	37	+ 6	49.5	50.2
St Thomas & Prince	.374	15	30	+15	65.5	63.0
Senegal	.178	32	14	−18	48.3	38.3
Seychelles	.740	2	2	0	70.0	89.0
Sierra Leone	.062	51	41	−10	42.0	20.7
Somalia	.088	44	46	+ 2	46.1	24.1
South Africa	.674	3	5	+ 2	61.7	70.0
Sudan	.157	39	25	−14	50.8	27.1
Swaziland	.458	9	10	+ 1	56.8	72.0
Tanzania	.268	25	48	+23	54.0	65.0
Togo	.218	29	24	− 5	54.0	43.3
Tunisia	.582	5	8	+ 3	66.7	65.3
Uganda	.192	30	45	+15	52.0	48.3
Zaire	.262	26	39	+13	53.0	71.8
Zambia	.315	19	23	+ 4	54.4	72.8
Zimbabwe	.397	13	16	+ 3	59.6	66.9

Sources: UNDP (1992) *Human Development Report 1992* and IBRD (1992) *World Bank Atlas 1992.*

72 Population growth rates 1960–90, 1980–90, 1990–2000; fertility rates 1970, 1990; death rate 1990

| | *Annual population growth rate* | | | *Fertility rate* | | *Death rate* |
	1960—90	*1980—90*	*1990—2000*	*1970*	*1990*	*1990*
Algeria	2.8	3.0	2.8	7.4	5.1	7.7
Angola	2.5	2.6	2.9	6.4	6.5	19.4
Benin	2.5	3.2	3.2	6.9	6.3	18.5
Botswana	3.4	3.4	3.4	6.9	4.7	10.5
Burkina Faso	2.4	2.6	3.0	6.4	6.5	17.8
Burundi	2.1	2.9	3.0	6.4	6.8	17.1
Cameroon	2.7	3.2	3.5	5.8	6.5	14.1
Cape Verde	2.1	2.6	3.4	7.5	5.4	7.5
Central African R.	2.3	2.7	3.0	4.9	5.8	17.1
Chad	2.1	2.4	2.6	6.0	6.0	18.7
Comoros	3.2	3.7	3.7	7.0	6.8	12.4
Congo	2.8	3.4	3.4	5.9	6.6	13.9
Djibouti	5.6	3.4	3.0	6.6	6.6	17.1
Egypt	2.4	2.5	2.0	5.9	4.1	10.0
Equatorial Guinea	1.1	2.0	2.6	5.0	5.5	18.8
Ethiopia	2.4	3.1	3.0	5.8	7.5	19.5
Gabon	3.0	3.6	3.2	4.2	5.8	16.4
Gambia	3.0	3.3	2.7	6.5	6.5	20.4
Ghana	2.7	3.4	3.2	6.7	6.2	12.5
Guinea	2.0	2.5	3.1	5.9	6.5	21.2
Guinea-Bissau	1.9	1.9	2.2	5.9	6.0	22.2
Ivory Coast	3.9	4.1	3.9	7.4	7.3	13.8
Kenya	3.6	3.9	3.8	8.0	6.6	10.5
Lesotho	2.4	2.7	2.9	5.8	5.6	11.7
Liberia	3.1	3.1	3.3	6.5	6.3	14.9
Libya	4.1	4.1	3.6	7.5	6.6	8.8
Madagascar	2.8	2.9	3.3	6.6	6.5	13.3
Malawi	3.1	3.4	3.6	7.8	7.6	19.8
Mali	2.5	2.5	3.2	6.5	7.0	19.9
Mauritania	2.4	2.4	2.9	6.5	6.8	18.2
Mauritius	1.7	1.0	1.0	3.6	1.9	6.3
Morocco	2.6	2.7	2.3	7.0	4.7	9.0
Mozambique	2.5	2.7	2.7	6.6	6.4	17.7

| | Annual population growth rate | | | Fertility rate | | Death rate |
	1960—90	1980—90	1990—2000	1970	1990	1990
Namibia	2.6	3.2	3.2	6.1	5.9	5.9
Niger	3.2	3.4	3.4	6.9	7.1	19.6
Nigeria	3.2	3.4	3.3	6.9	6.6	14.8
Rwanda	3.3	3.2	3.5	7.8	8.3	16.4
St Thomas & Prince	2.4	2.8	–	–	5.1	10.0
Senegal	2.8	3.0	2.9	6.5	6.5	16.9
Seychelles	–	0.7	–	–	2.8	8.0
Sierra Leone	2.1	2.4	2.7	6.5	6.5	22.5
Somalia	3.2	3.0	2.6	6.7	6.8	19.1
South Africa	2.4	2.4	2.2	5.7	4.2	9.3
Sudan	2.8	2.8	2.9	6.7	6.3	15.1
Swaziland	3.0	3.4	3.6	6.5	6.3	11.8
Tanzania	3.4	3.1	3.8	6.4	6.6	13.3
Togo	2.9	3.5	3.2	6.5	6.6	13.5
Tunisia	2.2	2.5	2.0	6.4	3.8	6.8
Uganda	3.6	3.2	3.7	6.9	7.3	14.9
Zaire	2.8	3.1	3.3	6.0	6.0	13.6
Zambia	3.4	3.7	3.8	6.7	6.7	13.0
Zimbabwe	3.2	3.4	3.1	7.7	4.9	9.6

Source: UNDP (1992) *Human Development Report 1992* and IBRD(1992) *World Bank Atlas 1992.*

73 Further reading: data sources and references

Brownlie, I. (1979) *African Boundaries: A Legal and Diplomatic Encyclopaedia*, London: Hurst.

Europa (Annual) *Africa South of the Sahara*, London: Europa.

Fage, J. (1978) *An Atlas of African History*, London: Arnold.

Grace, J. and Laffin, J. (1991) *Fontana Dictionary of Africa since 1960*, London: Fontana.

Hailey, Lord (1938) *An African Survey*, Oxford: Oxford University Press.

Hertslet, E., Sir (1909) *The Map of Africa by Treaty* (3rd edn), London: HMSO [Reprinted 1967: London, Frank Cass], 3 vols.

International Bank for Reconstruction and Development (World Bank) (1991) *World Development Report 1991: The Challenge of Development*, Oxford: Oxford University Press for the World Bank.

International Bank for Reconstruction and Development (World Bank) annual *World Bank Atlas*, Washington, DC: IBRD/WB.

Oliver, R. and Crowder, M. (eds) (1981) *The Cambridge Encyclopaedia of Africa*, Cambridge: Cambridge University Press.

Philip's (1992) *Geographical Digest 1992–93*, London: Heinemann–Philip Atlases.

Republic of South Africa (Annual) *South Africa: Official Yearbook of the Republic of South Africa*, Johannesburg: Chris van Rensburg.

South African Institute of Race Relations (SAIRR) (Annual) *Survey of Race Relations in South Africa*, Johannesburg: SAIRR.

United Nations Development Programme (UNDP) (annual from 1990) *Human Development Report*, New York and Oxford: Oxford University Press for UNDP.

United Nations, Food and Agriculture Organization (FAO) (1991) *Production Yearbook 1990*, Rome: FAO.

Williams, G. and Hackland, B. (1988) *The Dictionary of Contemporary Politics of Southern Africa*, London: Routledge.

Index

Note: figures in italics refer to map references; names in italics are former or alternative names; names in capitals are modern states of Africa